José Luis Reyes Carrillo
Pedro Cano Ríos
A. M. Reséndez Rubi M. Soto

Honey bees and melon pollination

José Luis Reyes Carrillo
Pedro Cano Ríos
A. M. Reséndez Rubi M. Soto

Honey bees and melon pollination

Imprint

Any brand names and product names mentioned in this book are subject to trademark, brand or patent protection and are trademarks or registered trademarks of their respective holders. The use of brand names, product names, common names, trade names, product descriptions etc. even without a particular marking in this work is in no way to be construed to mean that such names may be regarded as unrestricted in respect of trademark and brand protection legislation and could thus be used by anyone.

Cover image: www.ingimage.com

This book is a translation from the original published under ISBN 978-613-8-97694-3.

Publisher:
Sciencia Scripts
is a trademark of
Dodo Books Indian Ocean Ltd. and OmniScriptum S.R.L publishing group

120 High Road, East Finchley, London, N2 9ED, United Kingdom
Str. Armeneasca 28/1, office 1, Chisinau MD-2012, Republic of Moldova, Europe
Printed at: see last page
ISBN: 978-620-7-63573-3

Contents

Of the authors

Jose Luis Reyes Carrillo . Engineer Agronomist Zootechnician at the Institute of Technology and Studies Superiors of Monterrey (ITESM), Monterrey, Nuevo Leon; Master's degree in Administration in the ITESM Campus Laguna and Doctor in Sciences Agricultural by the Autonomous University Agrarian Antonio Narro (UAAAN). Member of the Health and Production Committee Apfcola of the Technical Council National Animal Health Advisor (SENASICA-SADER) and beekeeper since 1977. Member of the Academic Body Systems Sustainable for Production Agriculture (CASISUPA) and the Academic Innovation Network in Food and Sustainable Agriculture (RAIAAS). Competitor in 133 publications techniques and sciences , 8 books as author and co-author . Main advisor of 113 bachelor's theses , co -advisor of 132 and collaborator in 25 postgraduate theses . Professor-researcher at the Department of Biology of the UAAAN, Laguna Unit (UL) in Torreon, Coahuila, Mexico. National Researcher Level I of the National System of Researchers (SNI) of the National Council of Science and Technology (CONACYT)

Pedro Cano Rfos . Engineer Agronomist from the UAAAN in Saltillo, Coahuila in 1972, with a Master's degree (1977) and a Ph.D. in Genetics (1986) at New Mexico State University, USA. Researcher at the National Research Institute Forestry Agricultural and Livestock from 1972 to 2007. It belongs to CASISUPA. Has participated in the publication of 81 articles scientific , 90 articles in extenso, 10 books / pamphlets , 20 book chapters , 7 developments technological , 2 books edited and in the qualification of 318 bachelor's degrees , 27 masters in sciences and 19 doctorates . Currently performing as Professor-researcher at the Department of Horticulture at UAAAN-UL. Belongs to the SNI of CONACYT Level I with appointment emeritus as of December 31, 2027

Alejandro Moreno Resendez . Engineer Qtimico , Autonomous University of Coahuila (FCQ), Master in Sciences in Soils and Doctor in Sciences Agricultural , Autonomous University Agrarian Antonio Narro. Professor - researcher , Department of Soils , UAAAN-UL. Level I Researcher of the SNI-CONACYT. Member of the Mexican Society of Soil Science , AC Coordinator of the Academic Body Systems Sustainable for Production Agriculture , UAAAN-CA-14 and the Academic Innovation Network in Food and Sustainable Agriculture , sponsored by the State Council of Science and Technology of Coahuila and the Community of Higher Education Institutions of La Laguna (COECYT-CIESLAG). He has published 58 articles scientific , magazines national and international , 40 book chapters author and co-author . Advisor of 48 undergraduate theses and 15 postgraduate theses . Speaker at National and International Congresses with 15 conferences masterful . Member of the Arbitration Committee of 10 magazines periodic National and international .

Rubi Munoz Soto. Engineer Biochemistry graduated from the Autonomous University of Coahuila and master's degree in Human Capital Administration from CNCI University. From 2010 to 2014 head of the Program Engineering Academic in Processes Environmental and from 2014 to 2018 responsible for the School Control and University Services area. He has published 10 articles scientists and has participated as main advisor and collaborator in the graduation of 72 students from various careers in degree . At the moment professor-researcher and head of the Department of Biology at UAAAN-UL.

Prologue

In it world There are more than 20,000 species of bees and among them are many that pollinate , both the vegetation wild , like the crops of interest to humans . The best known bee is the bee meHfera (*Apis mellifera* L.) which is appreciated mostly for honey , pollen , wax , propolis , royal jelly and bee venom it produces. However the benefit of these products is insignificant compared to the function of these insects as pollinators , a service enormous environmental useful for the conservation of biodiversity and for the productivity of crops that benefit humanity . Through millions of years the bee meKfera evolved to become in a species surprisingly adaptable; development habits extremely gregarious and organized to live in big groups divided in a caste system compound by a Most infertile female bees , called workers , a group minority of fertile males acquaintances as drones and a bee queen . The workers perform all the hive jobs while the queen and the drones are the players . The main form of communication between all they are realized through the exchange of signals aromatic , that is , substances volatile chemicals . The signals that the queen emits maintain harmony and organization inside the colony, and everything this coupled with the diligent work of the workers , allows keep nails exceptional life conditions inside the hive .

In its path evolutionary bees meKferas developed a relationship symbiotic with plants angiosperms that would conquer the world . Are floors were developing nails structures attractive and colorful , which offer their visitors the reward of a substance delicious in sugars called nectar, in exchange for transporting their gametes masculine , - the pollen -, to others structures equal in others plants to be fertilized . This is how the flowers were formed . The pollen in itself constituted a reward , since it had a high content protein and plants produce it in big quantities , such way they could to satisfy the needs food of its visitors and even ensure that always had enough to carry to other flowers.

To the transport of pollen from a flower to another is called pollination , which is the mechanism that allows the sexual reproduction of plants and bees are experts pollinators . Additionally the processed nectar by bees it is the main constituent of honey . When after pollination is left over fertilization of the stigmata florals and with it the union of the gametes masculine and feminine , this gives rise to the fruits and seeds that later They could be dispersed to go to a fertile environment in where will germinate to become in new floors . This continuity It guarantees the permanence of the species , an evolutionary principle of survival .

Pollination through bees ensures the species vegetables a fertilization crossed , that is, between individuals different , thereby avoiding self -fertilization , which is degenerative for most species . This benefit of the function bee pollinator multiplies when the vision is magnified at the ecosystem level and even more when displayed from the global perspective . For this reason, it is considered that the activities of bees impact in a positive way in all he world since all plant species They depend directly or indirectly on bees for their growth and reproduction . This activity has such transcendence that Maurice Maeterlinck, naturalist , winner of the Nobel Prize in Literature and writer of the book *The Life of the Bee* said " Remove the bee from the ground and in he same blow you suppress at least one hundred thousand plants that will not survive ."

This work constitutes a compendium , example and explanation of something that happens in most of the ecosystems, both natural and modified by man and which is essential for plant life , the pollination carried out by bees . Is a synthesis of various study cases about he melon cultivation (*Cucumis melo* L.). In each chapter of the present book have been captured the efforts by contribute all data class practical about pore pollination the rest important melon cultivation . This species enjoys great acceptance among consumers given the pleasant consistency of your pulp , the so sweet taste of the fruit and its biocompound content functional as vitamins and antioxidants . It has a great demand at the level world and its price can become very high in some countries where they do not have the conditions suitable for your production how is he case of Northern Europe and Japan . For him contrary , in countries Like Mexico, the opposite happens , that due to the supply excessive he price it gets so low that it harms the producers .

Do you have in his hands a extremely useful work . In it he meets all the information possible about he melon cultivation and its pollination so that both producers inexperienced as the experienced can benefit from your reading . The intention of its authors is clear that in consideration of all the topics that are discussed here , the farmers can manage their crops in a more productive and sustainable way . It is everyone 's responsibility take in realize that the crops do not exist isolated but they form part of an ecosystem in where all the components are parts important of your biodiversity and are needed some to others to be able function . In it , a species prosper only when the others species or components prosper also . This is crucial to understand because the activities of man affect the ecosystems , and when a species go dead oun component is destroyed , everything the rest of the ecosystem looks affected .

Dr. Jose Luis Garda Hernandez National
Researcher Level III of the SNI
Leader of the Academic Body Consolidated " Sustainable Agricultural Production "
Head of the Research Department of the Faculty of Agriculture and Zootechnics , Universidad
Juarez of the State of Durango, Mexico

Preface

As part of the countless relations ecological ones that have place within the ecosystems terrestrial , there is one of extreme importance that is given in a way symbiotic among certain animals and plants , pollination . The agents regulators of this natural mechanism are pollinators . The equilibrium within these relationships is achieved thanks to the fact that group of pollinators this integrated by a great variety of insects , birds and mammals . It is from this way that nature try ensure pollination .

Among all the pollinators , the insects occupy the first place , both in number as in diversity and the most important are, without a doubt , the bees , for his stake in the pollination of crops that feed millions of people humans and many animals domestic . Others important Pollinators are wasps , flies , butterflies, some species of mosquitoes and many species of beetles .

The bees They pollinate more than 70% of the approximately 100 species that humans cultivated for various purposes. In total it is estimated that bees They pollinate more than 25,000 species of flowering plants . Other fact Important is that 60% of the fruits and vegetables that we consume today they would disappear Yeah they will stop being pollinated by the bees .

Regarding the activity apfcola , under the conditions of a climate semi-desert as that of the Comarca Lagunera , the bees face several challenges annually to survive . Some of the difficulties are the climate extreme and scarcity of resources in certain times of the year or some years . For other part , the conversion of the land that originally was vegetation cover native in a current mosaic of crops agricultural is not the most suitable scenario to develop a healthy beekeeping in the region, on all because exist crops that are regularly sprayed with insecticides chemicals that affect seriously or kill the bees .

The melon is one of the products most important horricoles in the Comarca Lagunera , which also occupies the first place in production of this level cultivation national . Melon orchards provide a niche opportunity for bees , as they offer resources foodstuffs and time they need the bee for a pollination adequate . Miscellaneous studies they have demonstrated that the use of bees in he cultivation gives results magnificent increasing he performance by hectare as well as he fruit size . However, this information is usually stay in the scientific dissemination media nothing more and does not reach the main interested as the farmers and the beekeepers . TO This is because when the theme between them , exist many questions regarding bee management in the melon orchards .

Aware of the problem , the authors of this book they took on the task of gathering all the information relevant and inviting to experts in he topic to present it in a manner technique and practice in order to inform , clear doubts and guide horticulturists and beekeepers in he driving suitable for both melon and bees in he vegetable patch .

I invite the readers to enjoy this construction site I'm sure your knowledge will increase in he management of the garden and that they will benefit from the benefits that bees offer as pollinators in these agroecosystems .

Dr. Arturo Daniel Tijerina Chavez

Regional DirectorNorth Regional Research Center Center

National Research Institute Forestry , Agricultural and Livestock

Thanks

collaborative work of undergraduate , master's and doctoral students is a reason for gratitude . in the research work that gave origin to the results presented in each chapter

the Soil Laboratory staff of the Autonomous University Agraria Antonio Narro Laguna Unit, QFB Norma Lidia Rangel Carrillo, TQ Juan Carlos Mej^a Cruz and TQ Jose Silverio Alvarez Valadez for your invaluable support in he analysis and processing of samples research projects

To the QFB Ana Maria Mej^a Fernandez and As. Ex . Paulino Aguilar Marchand of the Biology Laboratory of the Autonomous University Agraria Antonio Narro Laguna Unit by his aid in he pollen processing and microscope photographs

To the QFB Vfctor Alcantar Rosales of the Investigation and Assistance Center in Technology and Design of the State of Jalisco campus Apodaca, Nuevo Leon by he pesticide analysis in honey and wax

To the Laguna Beekeepers Association for his narrow collaboration in problem detection parasites and loss of hives by pesticides

To the photographers who with opportunity captured the images that illustrate the different chapters

To the Produce Coahuila Foundation , AC for his support economic in the investigations initials in the pollination of melon with bees meHferas in the Laguna Region

Autonomous University Agrarian Antonio Narro by he financing to research projects

To MVZ Juan Ernesto Enemegio Marrero in his disinterested help in creating the glossary

Now Doctor Roberto Quintero Dominguez brilliant student at the University Science Center Biological and Agricultural Sciences of the University of Guadalajara for the exhaustive and selfless review of this construction site .

" Feel gratitude and not expressing it is like wrap a gift and not give it "

William Arthur Ward

Authors and collaborators of the chapters
Azucena Vargas-Valero. National Institute of Forestry , Agricultural and Livestock Research (INIFAP), Edzna Experimental Field , Campeche- Pocyaxun , Campeche. Email: azvalero@gmail.com
Eduardo Castro Martinez . INIFAP La Laguna Experimental Field, Matamoros, Coahuila (deceased).
George Maltos Buendia. INIFAP Experimental Field La Laguna, Matamoros, Coahuila. Mail: maltos.jorge@inifap.gob.mx
Jose Luis Galarza Mendoza. Technological Institute of Torreon, Anna municipality of Torreon, Coahuila. Mail: galarzajl@yahoo.com.mx
John Cabrera Kings. Technological Institute of Torreon, Anna municipality of Torreon, Coahuila. Coahuila: goat_kingsj@yahoo.com.mx
Luis Henry Moreno Alvarado. Advisor agricultural Mail: emoreno.s.stairs@gmail.com
Pablo Preciado-Rangel. autonomous University Agraria Antonio Narro, Laguna Unit, Torreon, Coahuila. Email: ppreciador@yahoo.com.mx
Urban Nava Camberos. Juarez University of the State of Durango, Gomez Palacio, Durango. Email: nava_cu@hotmail.com
Veronica Avila Rodriguez. Juarez University of the State of Durango, Gomez Palacio, Durango. Email: vavilar@gmail.com
Veronica Garda Mendoza. business consulting agricultural Mail: agronegociosrentablesmktg@gmail.com

Foraging bees in the flowers of the cat 's eye herb (Photography Jordan Hernandez Sanchez)

"If bees only collected nectar from perfect flowers, they could not produce neither "a single drop of honey "

Matshona Dhliwayo

" It seems to me , Sancho, that there is no proverb that is not true ,
because they are all sentences taken from it experience , mother of
sciences all "

Don Quixote

Presentation

To get the maximums yields , in most crops require the activity of insects that carry out pollination . When the pollinators are scarce or inefficient , the quantity and quality of fruits decreases . Between the pollinators of the crops agricultural , bees meKferas are the best known . These laborious insects are extremely efficient to pollinate and also they take advantage the nectar of flowers to produce Honey .

So that the reader can know more about these interesting insects and learn to appreciate and respect them , they include in this book topics like life inside the hive and the behavior of bees , flow of entry and exit of bees from the hive , capture of pollen , collection of others materials during the time of scarcity , Africanization and the Disappearing Hive Syndrome . All of them are factors that affect the survival of bees and therefore their availability and efficiency in pollination .

The cantaloupe, Chinese or reticulated melon , is one of the crops that require bees for their pollination , but they are few the producers who place hives in their crops or if they place them , they do so in a inadequate .

The use of the bee meKfera for pollination of melon poses several issues , among which the importance of the location of the hives stands out . These cannot be in inside the growing area because hinder the work irrigation and weed control manuals and there are he risk of bites . Thus It is always better locate them on the periphery . Other consideration important is to determine the most opportune times to install and remove them having in account the influence of the distance from the apiary on the production and quality of the fruit ; Furthermore , he access himself to the hives to manage them It should be easy .

Although it is well known that the melon flower is attractive to bees , mainly by his pollen that is very abundant , and its nectar, although This is scarce , farmers need more information about he bee behavior during melon pollination : patterns and quantity of pollen capture , the temporal and spatial distribution of bees On his collection visits , the number optimal hives by hectare for one correct pollination and the bee foraging in floors wild or cultivated different from the melon plants in the vicinity of the crop . Due to the above, the fundamental purpose of this work is to respond to those questions and provide information so that farmer , with the support of the beekeepers , achieve maximum quality and performance in their fields , considering their time the control of weeds , pests and diseases that affect the crop , the work agricultural , the soil and nutrition too could affect bees and their function as pollinators .

" Books are the bees that carry he pollen from a intelligence to another " James Russel Lowell

1. Pollination

Azucena Vargas-Valero, Jose Luis Reyes-Carrillo and Alejandro Moreno-Resendez

Introduction

Pollination is the transfer of pollen from stamens , part male of the flower , until stigma , part feminine of a flower to another of the same species . This way it starts he mechanism of fertilization , and the consequent production of fruits and seeds [1] . This process can be carried out by vectors Biotics as the animals , and abiotics as he wind and the water , although most flowering plants depend on water . animals , mainly insects , among The most numerous and efficient pollinators are bees , which are therefore excellence participate in this activity of great importance economical and ecological in agroecosystems [2] . It is such his relevance , that most of the foods that are consumed in he world They depend on pollination [3] . Plants that require pollination attract responsible for this process through showy flowers , offering two important rewards : nectar and pollen . The nectar, composed by sugars , amino acids , minerals and substances aromatics , it is the source of energy for some animals as bees , butterflies , hummingbirds and moths , while the pollen It represents a source of protein and lipids , bees at the same time time they feed on the nectar, collect it and transport it to the nest as source of food for their offspring [4] .

The plants and pollinators They carry millions of years in coevolution , constituting a mutualism ratio [5,6] However, it has recently been reported a decrease significant increase in the number of pollinators , which has generated great concern due to its environmental repercussions as economical [7] . Various research they have permitted determine that the causes of the "crisis of the pollinators " are: 1) the introduction of species exotic insects that compete by resources as nectar and pollen and that they carry parasites , 2) deforestation , 3) large monoculture extensions , 4) the use indiscriminate use of agrochemicals and 5) the climate change [8,9,10] .

Due to the above, in he present chapter highlights the importance of pollinators , such as bees , ants , wasps , flies , moths , beetles , hummingbirds and bats are also described some of its characteristics and behavior , as well as the factors that affect his diversity and abundance .

Coevolucion

Exists a relationship evolutionary between plants and pollinators that has resulted in result he improvement of the mechanism of sexual reproduction of plants , on the one hand , and on the other hand another , adaptation anatomical and behavioral pollinators . Sexual reproduction allowed plants to produce seeds and fruits , and facilitated the propagation and survival of the species that constitute them . he food for animals and humans [11] . The adaptations of the pollinators led to the existence of pollinators that are generalists and visit a wide variety of plants to obtain his food as well as pollinators specific people who visit a small number of floors [5] . Inside this complex relationship has been generated a strong competition between pollinators by he access to nectar and pollen , and between plants , to attract pollinators through the color and smell of the flowers.

pollinator relationship this regulated by four factors:1) the abundance of resources floral , 2) the availability of nesting environments , 3) the presence of predators and pathogens and 4) pesticides [12] . For another side , the disturbance of the ecosystems , habitat conversion and landscape homogenization 13 they have I come altering plant communities pollinated and the function of pollination , putting at serious risk of disappearance pollinators and service ecological pollination . In places where This has happened , if there is existed as a service free and abundant , it has become in a service scarce and expensive .

The importance of pollinators in the ecosystems

Importance of pollinators for vegetation wild

Pollination is a service that pollinators they provide to the ecosystem . Without her or without them , the ecosystem was seriously affected [14] ; Kremen and collaborators [15] indicate that 60 to 90% of the vegetation wild depends directly or indirectly from the insects pollinators for your reproduction . The natural habitat provides food sources alternatives and nesting sites for pollinators . This is of utmost importance , since the provides resources when the crops are not in flowering [12,16] . It has been shown that the natural area around agricultural fields increases the diversity of pollinators [17] .

The importance of pollinators for agriculture

The abundance and diversity of pollinators they assure a sustained provision of pollination services to a wide diversity of plants . To level worldwide , pollination contributes considerably to agricultural performance [18] ; However the growth excessive activity agriculture has caused that throughout the last decades are cut down the forests and are cleared indiscriminately thousands of hectares to give room for the call agriculture intensive . The disappearance of vegetation wild put in danger to many pollinator species , such as bees wild , generating the inability of ecosystems to generate he service environmental pollination not only for vegetation wild but also for the big ones crop extensions [19] . If the systems agricultural will remain confined to a small scale , and refrained from using agrochemicals , they could contribute to supporting the natural pollinators and improve plant reproduction [20] . A global estimate indicates that the production of more than 87.5% of the plants cultivated depends mainly from the pollinating insects [6.21] and a third of the crops agricultural depend on your help for pollination . To level world has been estimated the economic value of pollination at 265 billion dollars annually [22] . Therefore , to complement the activity of the local pollinators , farmers rent bee or bumblebee hives [15] . The producers agricultural of the The United States of America and Europe are the ones have clearly identified he benefit economics of pollination [22] .

Even though the cereals are pollinated mostly by he wind or self-pollinate , it has been shown that with the participation of insects pollinators , give better yields and higher quality [23] . The crops horticultural such as nuts , fruits , vegetables , oilseeds and some fodder used for livestock feeding depend exclusively from the insects pollinators , so a decrease in their populations it affects seriously production of these crops [24]

sunflower flower with a bee collecting he pollen that constitutes his food (Juan Cabrera

Reyes Photography)

Main pollinators

Between the more efficient pollinators are found bees , butterflies, flies , beetles , moths , wasps , ants , birds and bats , among others . Through their visits to the flowers they facilitate and improve the production of the foods and have an impact positive in environment helping to maintain biodiversity in ecosystems [18,25] . The following describes the behavior of the main pollinator groups : insects , hummingbirds and bats .

insects

The insects understand he largest and most diverse group in the animal kingdom . Pollination entomophilia , that is , the one carried out by insects , it is the oldest and most important category in he group of animals pollinators : the The most representative orders are Hymenoptera , D^ptera , Lepidoptera and Coleoptera [26] .

Bees, bumblebees , wasps and ants

In it order Hymenoptera are the bees , the bumblebees of the genus *Bombus* , ants and wasps . TO this cluster belong the species that have developed the highest social organization, like the bee meHfera and several species of the genus *Bombus* . However, the vast majority of bees are solitary and exist many that are kleptoparasites , that is , they steal he food than others species they have collected or produced [29.30] .

Little ants visiting the carrot flowers (Photography Jose Luis Reyes Carrillo)

Characteristics and behavior of the Hymenoptera

Hymenoptera are small , compared to butterflies and moths , and , beetles , which have representatives who manage to overcome 15 centimeters . They have a device oral equipped with jaws that they use to cut and carry the food - like he pollen - or with tongues to lick or suck the nectar They visit the flowers whose appearance and mechanisms biological attract about everything to the bees , These flowers have certain characteristics that insects identify as signs , such as a coloration in shades ranging from white to yellow going for its various nuances , or light blue , a very sweet smell fragrant , a symmetrical tubular corolla small in whose bottom lies the nectar, and they usually offer pollen in quantities that can be scarce or very abundant [1, 27] . The concentration of sugars in the nectar that they prefer the insects for your food is in average 31.2% for species diurnal with a short tongue - less than a centimeter - like bees and flies . For species that are also diurnal but They have a long tongue - more than one centimeter -, like the bumblebees and some bees large , is 32.7% [28] .

The great diversity of Hymenoptera implies that they can be find in a great diversity of habitats. Bees , bumblebees and wasps are the more abundant pollinators , it is common find them in almost any Place of the world in where exist angiosperms . The ants , for his part , they are more abundant In tropical and subtropical regions , they live in he soil and trees ; like some species of bees , form very large colonies numerous , highly organized and divided in castes in which each member has a specific function [29,30] .

The bees meliferas and the bumblebees of the genus *Bombus* they have developed a sophisticated communication system . First of all They have an entire body laboratory for synthesis and release controlled substances volatile chemicals that serve to transmit signs behavioral , but Additionally , they are capable of informing , through movements rhythmic sequenced complex information about sources of food and water , demonstrating a amazing association of data such as direction, distance and quality of the target as well as colors , smells and shapes of flowers [31] . Both the bees meKferas , like the bumblebees and a tribe of this same bee family calls Euglossinas , they are of habits foraging and gathering generalists he pollen of various sources floral . The specialization occurs more often in solitary bees [5,32] . Being a generalist influences in the accessibility to food sources , in the digestibility and in the variation of the nutritional content [5] of what is collected . long - tongued species They have the advantage of power drink nectar from the flowers of a greater number of plants , while those species that have a short tongue [33] can only pecorear in flowers with tube corollas shorts . The ants They are also generalists [34] .

Research about he behavior of the Hymenoptera

Main jatropha pollinators are in his most insects . Beetles , flies and bees collect nectar and pollen , while ants , wasps and butterflies only looking for the nectar [35] . The Bendejo plant is pollinated by ants , bees tapeworms and flies . Among them , flies little ones and the ants they came looking for nectar and pollinivores , bees and flies medium and large , they sought pollen [34] . For another Side , de la Pena and collaborators [36] They report that the pollinators of avocado flowers in Mexico and Central America in his Most of them are bees , followed by wasps , flies ,

escarabajos y murciélagos.

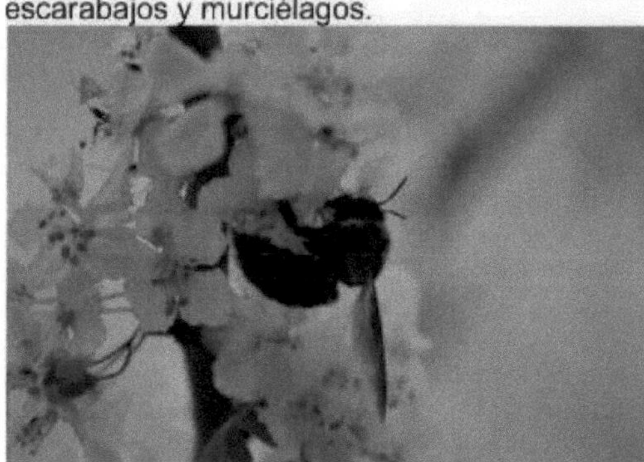

Wasp visiting the flowers of an American mesquite (Photography Isabel Blanco Cervantes)

In the pollination of the solitary palm , the greatest number of visitors floral They were from the beetles and Hymenoptera , however the stingless bees were the more efficient visitors [37] . Likewise way , the The most abundant and effective pollinator for cilantro was the bee [38,39] . In it cultivation of reticulated melon , pollination with bees guarantees the maximums returns when pollination is induced placing hives in cultivation [40] .

Flies and mosquitoes

exist approximately 150,000 species of d^pterans described at level worldwide , of which it is estimated that there are around 30,000 species in Mexico [26]. In it order Diptera are the flies , mosquitoes and mosquitoes , which although they are a group of insects very little appreciated , they play a role important in the interactions ecological as pollinators [41] , and are one of the three older and diverse animal groups of the world [42] .

Characteristics and behavior of the dipterans

Just like the hymenoptera , the d^pterans are generally of size small , though some flies they can reach up to seven centimeters in length . They present a device oral adapted to suck Kquidos . These insects They visit the flowers they have he design and mechanisms biological to attract about all flies : inconspicuous colors as he purple or green -, fragrance of substances in decomposition , scarce or absent nectar , little pollen and a Shallow corolla that is not symmetrical [27] .

Fly feeding on a cempoal flower (Photography Isabel Blanco Cervantes)

They live in a large number of terrestrial habitats , with greater diversity in the tropics . They are found near food sources , such as vegetation in he case of the d^pteros nectarivores and pollinivores , the matter in decomposition in he case of the necrophages and scavengers , and animals in he case of the d^pteros hematophages , which feed on the blood they suck from them . Flies and mosquitoes are essential in the chains trophic of the ecosystems since due to the diversity of their feeding habits occupy a important variety of niches ecological , mainly for being a source of food for other species [41] , as well as for your services as pollinators generalists highly uniform and effective [33] .

Lizard flower , succulent ornamental plant with flies attracted by his fetid smell (Photography Jose Luis Reyes Carrillo)

Research about he behavior of the dipterans

Flies are pollinators of at least 555 species wild and more than 100 species cultivated like mango , cashew , cocoa , onion , strawberry , broccoli , mustard , carrot , apple and cassava [42] . The flies Calliphoridae reproduce commercially as pollinators of the crops mentioned and some others as turnip , sunflower, buckwheat , garlic, lettuce , chili , and tomato [19] .

Mango flowers in a plantation in Culiacan, Sinaloa (Photography Hector Genaro Galindo Rodriguez)

Butterflies

Butterflies and moths - palomillas - belong to the order of lepidoptera . Its name is derived from the Greek " *lepidos* " which means scales and " *pteron"* wings, that is , "wings with scales ". As is usual in the nomenclature of beings alive , this name is very appropriate since indeed , when you see them under the microscope , you can notice that the wings are constituted by a delicate movie translucent deck by colored scales . Live in all the continents except in Antarctica and are more numerous and diverse in the tropics . exist approximately 150,000 species of butterflies and moths in he world , of which about 18,000 species They correspond to butterflies and the rest to moths . In Mexico they live approximately 1,800 species [26] .

Butterfly visiting the flowers of the Lantana plant (Photographs Isabel Blanco Cervantes)

Characteristics and behavior of the lepidoptera

These insects they visit the lepidopterophyllous flowers , I finish generic for all species morphologically adapted to pollination by lepidoptera . They are subdivided in diurnal butterflies , sphynx butterflies and nocturnal butterflies . Among its interesting adaptations are the flowers opening synchronized at night , corollas white to cream , corolla tube very narrow and long with a radial or bilateral symmetry , sweet fragrance and aroma emission that begins soft and becomes more intense in the afternoon , presence of abundant and hidden nectar deep in the tube , with a regular amount of pollen [27] . These insects They can be from very small to very large , with measurements from 20 millimeters to 30 centimeters from tip to tip of the wing. The antennas in butterflies they are longer , while in moths in his Most of them are shorter and thinner ; They have a device oral elongated , tongue-shaped , called spiritrompa , which normally this rolled up but they stretch to suck Liquids [26] .

Most moths are crepuscular or nocturnal , while butterflies can be diurnal or nocturnal . The life span of butterflies and moths is short , approximately one month , although some butterflies such as the Monarch can live up to nine months [32] .

Monarch butterfly visiting some Aster flowers (photography Jose Omar Enriquez Santacruz)

The Lepidoptera adults feed mainly from nectar. They transport incidentally he pollen that sticks to your body while They drink from the flowers and carry it in a flower to other contributing to this way to pollination . Regarding concentration average nectar sugars preferred by species nighttime is 44.6%, being the highest compared to the others pollinators (28); during he pecoreo feed on different sources floral as part of social learning [31] . There are generalist butterflies , which visit all type of flowers and others are specialists who visit only the flowers of certain plant species [43] . It has been shown that butterflies, nymphalids and moths

16

daytime sphinx colibri are able to identify and differentiate the color and size of the flowers [31] . The moths show a specialization in he pecoreo of equal or lesser as bees and birds [32] .

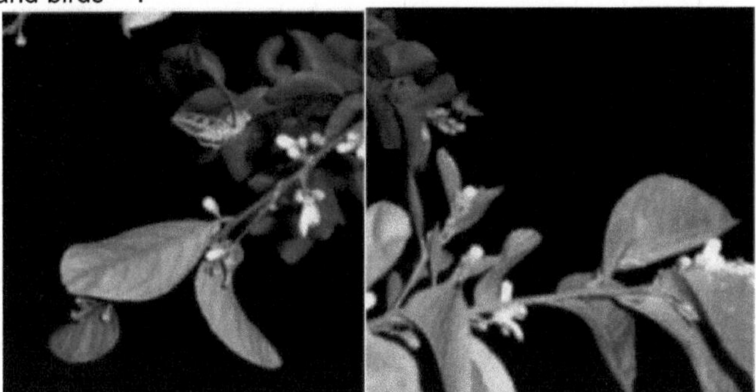

a moth sipping nectar from the flowers of a lime tree (Photography Jose Luis Reyes Carrillo)

Research about he behavior of the lepidoptera Main mustard pollinators , sesame and some vegetables are butterflies [44] . It has also been observed that they visit actively the macadamia crops , peanut , cashew and papaya [19] . Regarding the effectiveness pollinator of butterflies [45] it is known that although they are the most frequent visitors of some species , they are not the most important pollinators , because they do not transport neither deposit big pollen quantities ; by For example , when visiting the flowers of the wild plant " trumpet pine gilt " the butterflies were the most frequent visitors although were the less efficient in pollination [46] .

It has been discovered that there is a relationship proportionally direct between larger tongue butterflies , plants with greater length of the floral tube and greater sugar content of the nectar. Regarding the foraging habits , butterflies have a degree of specialization equal to or smaller than bees and hummingbirds [32] .

In order to preserve the pollinators in forests in North America, Hanula et al . [47] mention the importance of good forest management practices so that the plant species that already live there benefit time be pollinator reservoirs for recolonization of surrounding habitats .

Beetles

To level world are known around 358,000 species of beetles of the Order Coleoptera, grouped in 165 families . In Mexico there have been 114 families have been reported and more than 35,500 species are estimated [26] and it is the oldest and richest order in species . These pollinators influenced in a way important in the evolution of angiosperm flowers . Understands species of habits daytime as well as at night and in his majority , the Beetles are solitary animals [48] .

Characteristics and behavior of the beetles

The body of the adults is hard and varies in size well exist species that measure from a fraction of a miKmeter up to 15 centimeters long. Its cuticle or exoskeleton this compound by chitin and scleroproteins that give it rigidity . They have two pairs of wings; the first of them , called elytra , has lost his function as a wing and has become in a ngida structure protective of the other pair of wings that are functional . On the head they have a pair of antennae , a pair of eyes compounds that are only missing in some underground and cave-dwelling species and an apparatus oral chewer [26] equipped with a mandfoula that in many cases is extremely powerful . These insects They visit the cantharophilous flowers , this is with opening diurnal and nocturnal , with corollas of shallow , symmetrical tubes , of inconspicuous color , with a fruity fragrance in state of decomposition , without nectar but with abundant pollen [27] .

Most of the Coleoptera feed on different parts of plants , such as roots , stems , foliage , pollen , fruits or seeds . The beetles that pollinate flowers have mouthparts adapted for collection and nectar consumption , but about all pollen . The evolution of the beetles pollinators represents the adaptation of a type system generalist to one of a kind specialist well in the first no matter the opening nocturnal and are more prone to use color patterns in most flowers . The beetles pollinators they go to the flowers for three reasons: 1) for the floral reward of nectar, 2) in search for a mating site and 3) in search for protection since the flower offers protection against variations and inclement weather [37.49] .

Beetle feeding on pollen in a hediondilla flower (photograph Jose Omar Enriquez Santacruz)

Research about he behavior of the beetles

The beetles are the main family pollinators Annonacea , whose flowers, curiously , do not produce nectar. The petals are green , and have a fruity smell . in state of decomposition that also attracts thrips , flies and cockroaches, among other visitors [36,50,51] .

Tamaulipas Magnolia tree It has a pollination system specialized since you need necessarily the visit of the beetles of the genus *Cyclocephala* . The flowers offer as rewards a nectar high in carbohydrates and low fiber . They open in a night and are open 24 hours . The petals are thermal , that is, when they open present a temperature high that gradually is decreasing [52] .

Birds

The birds nectariferous by excellence are the hummingbirds are distributed exclusively in he American continent with around 330 species of which 57 live in Mexico [53] . However there are many others species of birds that are also nectariferous in greater or lesser extent .

Characteristics and behavior of birds

The birds pollinators They visit the flowers called ornithophiles , that is , those that have adaptations to be attracted by they . Characterized by have a opening diurnal , corolla or inflorescence branch predominantly red , yellow, blue or violet , without aroma, with abundant presence of nectar and regular amount of pollen . Its corollas they usually have long and narrow tubes , with bilateral symmetry [27] . All species of hummingbirds are small ; your weight goes from 2 to 24 grams , they have long, thin beaks , long , tubular , extensible tongues , and all species are nectarivorous [27] , [53] .

Hummingbird feeding on the ocotillo flower (Photography Isabel Blanco Cervantes)
The hummingbirds they can live in a wide variety of ecosystems , such as coastal areas , arid areas , jungles humid and dry , temperate and montane forests , and even in urban areas as parks and gardens . Are only missing in areas with climates very cold Some species perform long migrations distance , for example , in North America he Dusky buzzard breeds in Alaska and Canada and passes he winter in he central and southern Mexico. Are tiny Birds feed on the nectar of a wide variety of plants , pollinating the flowers of more than 1,000 species . Pollination occurs when the beak in the flower and spread his tongue to extract the nectar At that moment Their head and neck are impregnated with the pollen of the stamens and subsequently transports it to the pistils of other flowers. How are you relationship is beneficial for both the plant and the hummingbirds , it is identified as a mutualistic interaction [53] .

The sugar concentration of the nectar that they prefer for their diet [28] is 27%. Species of beaked hummingbirds short show greater specialization in his eating habit [32] , [54] attracted for the red flowers tubular , which produce quantities moderate nectar and sugar concentration in he range favorite by these birds [55] .

Research about he behavior of the hummingbirds

In a study in the Valley of Puebla on the pollination of the columnar cacti , known as " organs " it was found that during he d^a its flowers are visited by hummingbirds and during the night by bats . The accumulation average daily nectar in the flowers of these cacti was 0.087 microliters , with a sugar concentration from 25 to 48%. In it d^a were observed three species of hummingbirds and at night , two species of bats diverse moths . The hummingbirds presented greater individual visit frequencies pollen carriers compared to bats [55] .

In the eastern Colombian mountain range , it was found that visitors passion fruit floral were the stingless bees , the hummingbirds , butterflies, wasps and flies , however, for his behavior and frequency of visits are last ones don't have an impact significant in pollination so they are considered only as visitors floral , not as pollinators [56] .

Murcielagos

The chiropterans are those mammals winged ones we know more commonly as bats . In it world have been described 927 species and in Mexico there are around 137 species . The number of species It tends to be higher in tropical regions . Most bat species Mexicans are insectivores although include others little ones invertebrates in his diet (67.88%), but There are also bats that feed on fruits (16.06%), nectar and pollen (8.76%), carnivores (3.65%), blood (2.19%) and fish (1.46%) [57] . Those that feed on nectar and pollen have adapted to obtain his food

19

of some flowers and in their visits , just as happens with the others groups of pollinators that we have seen, carry he pollen from a flower to another thus allowing pollination .

Characteristics and behavior of the bats

They visit the flowers they have adaptations to receive the visit of the bats and being pollinated by they ; they are known as species chiropterophila . Among its characteristics are the flowers with opening nocturnal , non- flashy color as green and white , a mushroom smell that is more intense during the night , abundant nectar and pollen , corolla shaped like a large , pendulous bell [27] .

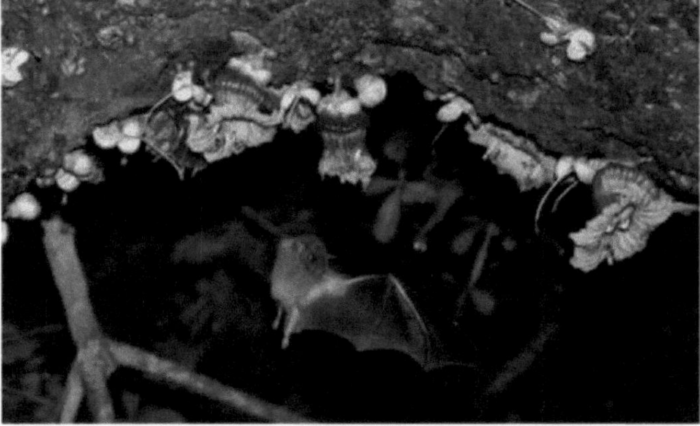

tecomate tree in flower visited by bat in Urique , Chihuahua (photography Javier Cruz Nieto)

The bats prefer for your nectar feeding of 44.6% sugar . This concentration is higher compared to what most people prefer . diurnal pollinators [28] . The flowers differ from those pollinated from d^ to in terms of floral anthesis , flower color and size , odor and nectar volume [58] . For him flight night , the bats they navigate with a system specialized echolocation , emitting ultrasound pulses shorts that allow them recreate in his brain a perfect map of your environment and locate objects and prey from the echoes that return . This way they move smoothly in absence of light [59] .

There are two families of bats specialized in flowers. The first is constituted by the bats found in Africa, Asia and Europe. They are considered as visitors flower opportunists because his diet is very diverse and among others food also They consume nectar. The second is integrated by the bats found In America. They are nectarivores specialized well they feed exclusively from the floral resources [58.60] .

Not like others pollinators as insects and birds , bats transport big amounts of pollen in their bodies and transfer it to large distances . Plants that pollinate they usually meet in type habitats arid and semi-arid with abundance of cacti and agaves, as well as in tropical forests both dry as humid with species as cultivated eucalyptus and banana trees [58] .

Research about he behavior of the bats

Muchhala et al. [61] They found that the flowers of the Afelandra were pollinated by bats that carried big quantities of pollen and good quality , being the responsible for approximately 70 % of your pollination ; by other side , there is a specialization of bats , due to the absence of other plant species and pollen volume that these they transfer in trees tropical comes to transport big quantities he pollen - up to 18 kilometers away - and are the main pollinators of agaves and epiphytes , which are plants that grow about another vegetable [58] .

Others pollinators

exist others animals that can get to pollinate some plant species , among them are the marsupials , various species of monkeys and rodents , the lemur, the squirrel arboricola and some species of parrots , which they use the nectar like food and at the same time perform he paper secondary pollination [18,31,62] . In all these cases he pollination success It depends on the characteristics of the animal, such as size , life cycle , way of foraging , requirements nutritional and habitat, and of the plant, as size , shape of the flower , reward offered , flowering time . Others factors that can influence are the climate and the interactions between the different pollinator species [63] .

Factors that affect the diversity and abundance of pollinators

At a global level, many pollinators are found currently under threat of extinction due to the impact of the activities human 40% of them this represented by bees and butterflies and 16.5 % by birds and bats 18 ' There are multiple factors that affect the diversity and abundance of pollinators but among the main ones are he change in land use , use indiscriminate use of agrochemicals poisonous and effects related to the climate change [64] .

Land - use change

The modification of the landscape through the fragmentation, degradation and destruction of natural habitats for the establishment of new ones . development spaces urban , practices agricultural intensive and monocultures they have injured to a degree superlative plant- pollinator interactions at the individual , population and community scale [15.65] in many ecosystems of the world .

Dismantling of a mosque to plant fodder in the Lagunera Region (Photographs Samuel Atahualpa Ramirez Macias)

Under these pressures , the more generalist pollinators have greater chances of surviving . The bees meKferas , by example , by having habits food polyfloral and the ability to visit communities vegetables diverse , present a minor vulnerability and greater resistance . For his part , the pollinators that have subsistence allowance monoflorals are found in greater danger and this shows because the land use change this affecting seriously to pollinators [66] . However the land use change not necessarily represents something negative . If the human being will respect the ecosystems when carries out its activities productive and will carry them out without destroying the natural resources being inserted in a way discreet and intelligent in the wild , the availability , diversity and abundance of nectar and pollen resources would follow providing a nutrition suitable for everyone the pollinators , which after all are the responsible for the production of most of our food .

Use of agrochemicals

Pollinators are dying poisoned by he use indiscriminate use of agrochemicals [67] , [68] . The pesticides used in the agriculture systems intensive not only kill pests but they are also lethal for many others beings alive . For the bees , in he best of the scenarios , cause serious disorders in his foraging capacity [69] . To level world the spotlights red are on from does three decades throughout which it comes warning to the producers and the population in general about the harmful effects that pesticides have about the insects pollinators and in particular on bees [18,70,71] .

dead bees at the entrance to the hive by poisoning with a pesticide (Photography Octavio Vazquez Calvete)

Climate change

The change climatic is recognized worldwide as one of the greatest threats to biodiversity . High temperatures, droughts , and floods , among others events climatic extremes , have caused changes phenological and imbalances between plant and pollinator populations that subsequently they can lead to the extinction of both. Although, the change climate and habitat destruction have been related to the global decline in biodiversity [18.72] . This generates changes in the biodiversity of species according to altitudes, which leads to a mismatch space between plants and pollinators ; has been shown in the butterflies that changes altitudinal in response to climate decrease species richness associated with disturbances , with reductions severe where habitat destruction is greatest [73] .

It is anticipated that the global warming will increase the events climatic extremes , like storms , floods and droughts . This can have a great impact in local pollinator communities [65] ; there will be displacements in flowering patterns , whether they come early or late and lose synchronicity with the arrival or the birth of their natural pollinators , losing these his food source . It is estimated that between 17 and 50 % of species pollinators they will suffer food shortage due to disturbances in the flowering patterns of plants [74] .

Burn intentional mesquitera to make charcoal and firewood (Photographs Jose Luis Galarza Mendoza)

Conclusions

Among the great diversity of pollinators are bees , wasps , ants , flies , butterflies, beetles , moths , birds and bats . Its diversity is result of a narrow coevolution between plants and animals that pollinate them . Pollination is key to the continuity of plant species and therefore their diversity . Lose heterogeneity vegetative would put in risk he heritage biological of the planet so it is no longer optional but imperative ban he use of poisons agrochemicals that affect directly or indirectly to the pollinators , so as to diversify the crops in agricultural areas , in order to increase his abundance and wealth . The decrease in pollinators would not only cause an agricultural crisis in the crops dependent on them , but the consequences ecological they will be greater well they are the base of the chains trophic of the ecosystems and are the responsible for the quantity and quality of the production of many of our food .

mesquites survivors of clearing for fodder planting (Photography Samuel Atahualpa Ramfrez Madas)

References

1.   Buchmann SL, Nabhan GP. The forgotten Pollinators. Island Press/Shearwater Books, Washington DC;1996;292p.
2.   Kelly D, Sork VL. Mast Seeding in Perennial Plants: Why, How, Where? Annu Rev Ecol Syst. 2002;33:427 - 47.
3.   Food and Agriculture Organization of the United Nations (FAO). Principles and advances about pollination as service environmental for agriculture sustainable in countries of Latin America and the Caribbean. 2014. [Online]: http://www.fao.org/3/a-i3547s.pdf (Consulted 09/12/20).
4.   Vaudo AD, Tooker JF, Grozinger CM, Patch HM. Bee nutrition and floral resource restoration. Curr Opin Insect Sci. 2015;10:133-41.
5.   Nicolson SW, Wright GA. Plant - pollinator interactions and threats to pollination: perspectives from the flower to the landscape. Functional Ecol. 2017;31:22-5.
6.   Ollerton J. Pollinator Diversity: distribution, ecological function and conservation. Annu Rev Ecol Evol Sist. 2017;48:353-76.
7.   Gallai N, Salles JM, Settele J, Vaissiere BE. Economic valuation of the vulnerability of world agriculture confronted with pollinator decline. Ecol Econ. 2009;68(3):810-21.
8.   Tur C. Changes in land use as responsible for the decline of pollinators . Ecosystems . 2017;27(2):23-33.
9.   Bartomeus I, Bosch J. Loss of pollinators : evidence , causes and consequences . Ecosystems . 2018;27(2)(2):1-2.
10.   Sosenski P, Dominguez CA. The value of pollination and risks you face as service ecosystem . Rev Mex Biodivers . 2018;89:961 -70.
11.   Nicholls CI, Altieri MA. Plant biodiversity enhances bees and other insect pollinators in agroecosystems. A review. Agron Sustain Dev. 2012;33:257 .
12.   Minarro M, Garcia D, MaMnez -Sastre R. Insects pollinators in agriculture : importance and management of its biodiversity . Ecosystems . 2018;27(2):81-90.
13.   Kovacs-Hostyanszki A, Espundola A, Vanbergen AJ, Settele J, Kremen C, Dicks LV. Ecological intensification to mitigate impacts of conventional intensive land use on pollinators and pollination. Ecol Lett. 2017;20:673-89.
14.   Garantonakis N, Varikou K, Birouraki A, Edwards M, Kalliakaki V, Andrinopoulos F. Comparing the pollination services of honey bees and wild bees in a watermelon field. Sci Hortic. 2016;204:138-44.
15.   Kremen, C., Williams NM, Aizen MA, Gemmill-Herren B, LeBuhn G, Minckley R, et al. Pollination and other ecosystem services produced by mobile organisms: a conceptual framework for the effects of land-use change.

Ecol Lett. 2007;10(4):299-314.

16. Simba LD, Foord SH, Thebault E, Van VFJ F, Joseph GS, Seymour CL. Indirect interactions between crops and natural vegetation through flower visitors: the importance of temporal as well as spatial spillover. Agric Ecosyst Environ. 2018;253:148-56.

17. Alomar D, Gonzalez-Estevez MA, Traveset A, Lazaro A. The intertwined effects of natural vegetation, local flower community, and pollinator diversity on the production of almond trees. Agric Ecosyst Environ. 2018;264:34-43.

18. Food and Agriculture Organization of the United Nations (FAO). The importance of bees and other pollinators for food and agriculture. Zirovnica, -Republic of Slovenia- Ministry of Agriculture, Forestry and Food. 2018. p.16 [En linea]: http://www.fao.org/documents/card/en/cZi9527enZ (Consulta 10/09/20).

19. Klein AM, Vaissiere BE, Cane JH, Steffan-Dewenter I, Cunningham SA, Kremen C, et al. Importance of pollinators in changing landscapes for world crops. Proc R Soc B. 2007;274:303-13.

20. Hass AL, Kormann UG, Tscharntke T, Clough Y, Fahrig L, Martin JL, et al. Landscape configurational heterogeneity by small-scale agriculture, not crop diversity, maintains pollinators and plant reproduction in western Europe. Proc Roy Soc B. 2018;285(20172242).

21. Ollerton J, Winfree R, Tarrant S. How many flowering plants are pollinated by animals ? Oikos. 2011;(321):321-26.

22. Lautenbach, S, Seppelt R, Liebscher J, Dormann CF. Spatial and temporal trends of global pollination benefit. Plos One. 2012;7(4).

23. Nicole W. Pollinator power: nutrition security benefits of an ecosystem service. Environ Health Perspect. 2015;123(8):A210-A215.

24. Spivak M, Mader E, Vaughan M, Euliss NH. The plight of the bees. Environ Sci Technol. 2011;45(1):34-8.

25. Rader R, Bartomeus I, Garibaldi LA, Garratt MPD, Howlett GB, Winfree R. et al. Non-bee insects are important
contributors to global crop pollination. Proc Natl Acad Sci USA. 2016;113(1):146-51.

26. Llorente- Bousquets J, Ocegueda S. State of knowledge of the biota, *In* Natural capital of Mexico. Current knowledge of biodiversity . Conabio , Mexico. 2008;1:283–322 .

27. Wolff D. Nectar sugar composition and volumes of 47 species of Gentianales from a southern Ecuadorian montane forest. Ann Bot. 2006;767–77.

28. Chalcoff VR, Aizen MA, Galetto L. Nectar concentration and composition of 26 species from the temperate forest of South America. Ann Bot. 2006;97: 413-2

29. Reyes-Novelo DG, Melendez- Ramurez A. Wild bees (Hymenoptera: Apoidea) as bioindicators in the neotropics. Trop Subtrop Agroecosystems. 2009;10:1 -13.

30. Sharkey NA, Nieves- Aldrey JL. Hymenoptera . *In* : The tree of life : systematics and evolution of the beings alive , Edition : Impulso Global Solutions, SA Madrid, Spain. Vargas P, Zardoya R. (eds.) 2012;322-33.

31. Jones PL, Agrawal AA. Learning in insect pollinators and herbivores. Annu Rev Entomol. 2017;(62):53-71.

32. Waser NM, Ollerton J (eds.). Plant-pollinator interactions: from specialization to generalization. The University of Chicago Press, Chicago, USA. 2006;445p.

33. Amorim FW, Haber WA, Johnson SD, More M, Frankie GW, Stanley DA, et al. The long and the short of it: a global analysis of hawkmoth pollination niches and interaction networks. Functional Ecol. 2017;31(1):101-15.

34. Gomez JM, Zamora R. Generalization vs specialization in the pollination system of *Hormathophylla spinosa* (Cruciferae). Ecology. 1999;80(3):796-805.

35. Samra S, Samocha Y, Eisikowitch DVY. Can ants equal honeybees as effective pollinators of the energy crop *Jatropha curcas* L. under Mediterranean conditions? Global Chang Biol. 2014;6:756-67.

36. De la Pena L, Perez V, Alcaraz L, Lora J, Larranaga N, Hormaza I. Pollinators and pollination in fruit trees subtropics : implications in management , conservation and food safety . Ecosystems . 2018;27(2):91-101.

37. Nunez-Avellaneda LA, Carreno JI. Pollination by bees in *Syagrus orinocensis* (Arecaceae) in the Orinoquia Colombian . Acta Biol Colomb. 2017;22(2):221-33.

38. Chaudhary OP, Singh J. Diversity, temporal abundance, foraging behavior of floral visitors and effect of different modes of pollination on coriander (*Coriandrum sativum* L.). J Spices Aromat Crop. 2007;16(1):8-14.

39. Painkra GP. Foraging behaviour of honey bees on coriander (*Coriandrum sativum* L.) flowers in Ambikapur of Chhattisgarh. J Entomol Zool Stud. 2019;7(1):548-50.

40. Reyes-Carrillo JL, Galarza-Mendoza JL, Munoz-Soto R, Moreno-Resendez A. Territorial and spatial diagnosis of beekeeping in the systems agroecological of the Lagunera region . Rev. Mexicana Cien Agnc . 2014;5 (2): 215 - 28.

41. Skevington JH, Dang PT. Exploring the diversity of flies (Diptera). Biodiversity. 2002;3(4):3-27.

42. Ssymank A, Kearns C, Pape TC. Pollinating Flies (Diptera): A Major Contribution to Plant Diversity and Agricultural Production. Biodiversity. 2008;9:86-9 .

43. Tobar LD. The pohnic loads in the butterflies of the basin of the no Oak- Quindfo . Caldasia 2001;23(2):549-57.

44. Bhaskar JD, Kaushik P, Nabanita B, Wine KB. Diversity study of insect pollinators of cultivated rabi crops in surrounding areas of Barpeta town in Assam, India. Rev. Inter J Res. 2018;5(4):172-7

45. Barrios B, Pena SR, Salas A, Koptur S. Butterflies visit more frequently, but bees are better pollinators: The importance of mouthpart dimensions in effective pollen removal and deposition. AoB Plants. 2016;8:plw001.

46. Krauss J, Steffam-Dewenter I, Tscharntke T. How does landscape context contribute to effects of habitat fragmentation on diversity and population density of butterflies? J Biogeography. 2003;(30):889-900.

47. Hanula JL, Ulyshen MD, Horn S. Conserving pollinators in North American forests: a review. Nat Areas J. 2016;36(4): 427-39.

48. Bernhardt P. Convergent evolution and adaptive radiation of beetle-pollinated angiosperms, Plant Syst Evol. 2000;222:293-320.

49. Li, JK Huang SQ. Effective pollinators of Asian sacred lotus (*Nelumbo nucifera*): contemporary pollinators may not reflect the historical pollination syndrome. Ann Bot. 2009;104(5): 845-51.

50. Caleca V, Lo Verde G, Ragusa S, Tsolakis H. Insect and hand pollination of *Annona* spp in Sicily. Phytophaga. 2002:117-27.

51. Silva-Costa M, Silva-Ricardo J, Paulino-Neto HF, Barbosa-Pereira MJ. Beetle pollination and flowering rhythm of *Annona coriacea* Mart. (Annonaceae) in Brazilian cerrado: Behavioral features of its principal pollinators. PLos One 2017;12(2):e0171092.

52. Dieringer G, Cabrera RL, Lara M, Loya L, Reyes-Castillo P. Beetle pollination and floral thermogenicity in *Magnolia tamaulipana* (Magnoliaceae). Int. J Plant Sci. 2009;160(1):64-71.

53. Arizmendi MC, Berlanga H. Colibnes of Mexico and North America . National Commission for the Knowledge and Use of Biodiversity (CONABIO). Mexico, DF 2014;160p.

54. LoPresti EF, Goidell J, Mola JM, Page ML, Specht CD, Stuligross C, et al. A lever action hypothesis for pendulous hummingbird flowers: experimental evidence from a columbine. Annals of Botany. 2019;125(1):59-6

55. Dar S, del Coro-Arizmendi M, Valiente-Banuet A. Diurnal and nocturnal pollination of *Marginatocereus marginatus* (Pachycereeae : Cactaceae) in Central Mexico. Ann Bot. 2006;97: 423-2

56. Medina-Gutierrez J, Ospina-Torres R, Nates-Parra G. Effects of altitudinal variation on pollination in purple passion fruit crops (*Passiflora edulis* f. edulis). Acta Biol Columbus. 2012;17(2):381-95.

57. Sanchez O. Bats of Mexico. CONABIO. Biodiversity . Mexico, DF 1998;20:1-11.

58. [PubMed] Fleming TH, Geiselman C, Kress WJ. The evolution of bat pollination: a phylogenetic perspective. Ann Bot.
2009;104(6):1017-43.

59. Jakobsen L, Olsen NM, Surlykke A. Dynamics of the echolocation beam during prey pursuit in aerial hawking bats. Proc Natl Acad Sci. 2015;1-6.

60. Stewart AB, Dudash MR. Differential pollen placement on an Old World nectar bat increases pollination efficiency. Ann Bot. 2017;117:145-52.

61. Muchhala N, Caiza A, Vizuete JC, Thomson JD. A generalized pollination system in the tropics: bats, birds and *Aphelandra acanthus*. Ann Bot. 2008;103:1481-87.

62. Wester P. The forgotten pollinators-first field evidence for nectar-feeding by primarily insectivorous elephant-shrews. J Poll Ecol. 2015;16(15):108-11.

63. Garibaldi LA, Morales LC, Ashworth L, Chacoff PN. Los polinizadores en la agricultura. Ciencia Hoy. 2012;21(126):35-42.

64. Barron AB. Death of the bee hive: Understanding the failure of an insect society. Curr Opin Insect Sci. 2015;10:45- 50.

65. Goulson D, Nicholls E, Botias C, Rotheray EL. Bee declines driven by combined stress from parasites, pesticides, and lack of flowers. Science. 2015;347(6229):1255957-8.

66. Alaux C, Ducloz F, Crauser D, Le Conte Y. Diet effects on honeybee immunocompetence. Biol Lett. 2010;6(4):562-65.

67. United Nations Environmental Program (UNEP). UNEP emerging issues global honey bee colony disorders and other threats to insect pollinators. 2010. [En Hnea]:https://www.unenvironment.org/es/node/12059 (Consulta 30/09/20).

68. Hladik ML, Vandever M, Smalling KL. Exposure of native bees foraging in an agricultural landscape to current-use pesticides. Sci Total Environ. 2016;54:469-77.

69. Desneux N, Decourtye A, Delpuech J. The sublethal effects of pesticides on beneficial arthropods. Annu Rev Entomol. 2007;52(1):81-106.

70. Cutler GC, Scott-Dupree CD. Exposure to clothianidin seed-treated canola has no long-term impact on honey bees. J Econom Entomol. 2007;100(3):765-72.

71. Chauzat MP, Carpentier P, Martel AC, Bougeard S, Cougoule N, Porta P, et al. Influence of pesticide residues on honey bee (Hymenoptera: Apidae) colony health in France. Environ Entomol. 2009;38(3):514-23.

72. Bellard C, Bertelsmeier C, Leadley P, Thuiller W. Courchamp F. Impacts of climate change on the future of biodiversity. Ecol Lett. 2012;15:365-77.

73. Forister ML, Mccall AC, Sanders NJ, Fordyce JA, Thorne JH, O'Brien J, et al. Compounded effects of climate change and habitat alteration shift patterns of butter fly diversity. Proc Natl Acad Sci USA. 2010;107(5):1-5.

74. Memmott J, Craze PG, Waser NM, Price MV. Global warming and the disruption of plant-pollinator interactions. Ecol Lett. 2007;10:1-8.

"For bees , the flower is the source of life . For flowers, the bee is the messenger of love."

Kahlil Gibran

2. Life inside the hive

Jose Luis Reyes-Carrillo

Social organization

bee colonies meHferas They have a caste system that involves different types of bees doing different tasks . First of all each colony has a queen Responsible for the activity reproductive inside the hive . In second place , the bee population meKferas has a small number of drones or male bees , in addition to that , they have a large number of bees workers that are sterile females [1] .

The bees workers are morphologically identical , but they can be pull apart According to their behavioral roles , they show a division of labor by age where the bees newly emerged from the cell After metamorphosis , they tend to perform tasks as he care of breeding and maintenance of the interior of the hive , and the bees workers greater perform tasks in the exterior, such as the search for food , be it pollen and nectar according to need , and substances essential for the colony such as he propolis and water [1-3] . Bee breeds can be distinguish by differences well-defined anatomical [4] .

Kingdom	Animal
Phylum	Arthropoda
Class	insect
Order	Hymenoptera
Family	Apidae
Gender	*Apis*
Species	*mellifera* L.

spider

Illustration of the different bee breeds melifera . The green dot on the back of the thorax it is used to identify the queen of the year 2019 within the colony according to the international queen marking color 5 (Photographs Jose Luis Reyes Carrillo)

Division of labour

The division of labor in bees it is one of the phenomena better explored in he study of animal behavior . Although the investigations date back to 1800s , the dedicated experimental work began in the 1930s , and has continued until the present with numerous laboratories of research centers that address he problem from all biological perspectives [4] .

In a bee colony meKferas , there are several phenomena in those who behavior of a queen and tens of thousands of her daughters workers they must coordinate to maintain survival and function . In established colonies , one of these phenomena is the reproductive division of labor , where the queen lays eggs that are cared for until age adult by bees workers optionally sterile , which generally they lay eggs only in absence of a queen and brood . The workers also exhibit a division of labor based in age , where they specialize in different tasks as they mature , with bees youths -" nurses "- who perform brood care and bees older ones - " pecoreadoras " - who collect environmental resources . The pecoreo in itself is another process essential at the colony level where the activities of a part substantial population are finely coordinated to facilitate he discovery efficient and profitable exploitation of food sources [6].

Bees foraging lavender flowers (Photographs Jose Luis Reyes Carrillo)

The Queen
The queens of the insects social they can impose cooperation and sterility on the workers of various ways , such as aggression physics and chemical communication [7-8] . in the bee meKfera , the queen uses the " queen 's mandibular pheromone ", which prevents the activation of the ovaries of worker bees [7] inhibiting oviposition ; also provokes a " courtship response " of workers , in which they meet in around the queen Forming an entourage , they touch her with their antennae and take care of her [9] . This pheromone is shared by the workers when exchanging food among them - this activity is called trophallaxis - and when grooming thus indicating to the entire colony the presence of the queen through the dissemination of these " smells " real ".

The bee queen followed by he entourage that touches it with the antennas (left) and cells with eggs in the Upica shape of a white sausage (right) (Photographs Jose Luis Reyes Carrillo)

The task of caring for the queen - feeding , grooming , examining and grooming her - is a form of cooperation. necessary for colony function among bees workers and the queen ; so the workers they must feed and maintain her since the queen this busy individually in the task of laying eggs; Therefore , the response of the workers to the queen is of key importance for the health of the colony. Exists a natural variation in response to the queen among the workers of a bee colony meliferas according to age [10-11] and this variation in response contributes to the division of labor in the colony, that is, the individuals specific are more likely to respond and therefore care for the queen [12] , so that the retinue of bees is replaced and renewed constantly .

worker bees

Cleaner bees . After the egg position and subsequent stadiums larvals , metamorphosis takes place , where at 21 days will emerge a worker adult The bees newly emerged they cannot fly or bite and are therefore immature in his development . The first days of the life of one bee pass developing and acquiring are skills . The repertoire of tasks during this period consists in cleaning the cells , with the rest of the time inactive or grooming that consists in the neat Cleaning the body , antennae and legs individually and collectively . This group of bees does not include a component functional critic of the colony , since the members of other castes also They clean the cells . Therefore, this breed It can be him result of restrictions physiological in he development of honey bees [4] and this occurs in the three first days of life as bee adult after emerging from the cell .

Cells operculated , that is, sealed with wax, worker where metamorphosis takes place (left image) and bee worker newly cell exit with hairs still colorless that cover it (right image) (Photographs Jose Luis Reyes Carrillo)

Nurse bees . bee activity nurse usually lasts approximately a week , from 4 to 12 days of life adult The workers wet nurses they feed the young larvae with a digested protein mass [of 4] and rich in lipids [5] called " pollen bread " composed of pollen , honey and various enzymes added for the bees that transform he product through a fermentation . Unlike pollen alone , as is the case of others bees social , presumably The " pollen bread " increases the growth rate of the larva because it does not have to digest the outer layer of pollen , which is very resistant . In addition to feeding the young , the nurses also transfer his secretion protein to younger and older bees in he nest and, therefore , are critical for its development and maintenance . The wet nurses also They take care of the queen by forming he entourage to his around , regulate his behavior through the speed at which they feed it and act as bees messengers by spreading the queen 's pheromones about the nest [4] reinforcing constantly the information of the queen 's presence through the " smells" real ".

packaged pollen in the cells , where you can see their different colors according to he floral origin and workers patrolling (left image) and workers wet nurses feeding the larvae in the breeding cells (right image) (Photographs Jose Luis Reyes Carrillo)

Medium bees age . Medium bees age remain in this category for a little more than one week , from 12 and up to 21 days ; are They have a repertoire of tasks distributed by all he nest , although his distribution overlaps that of the nurses , their behavior is quite different , since they do not show interest in breeding . Instead , his range of tasks understands about 15 activities different that range from the construction and maintenance of cells , the reception and processing of nectar, to the protection of the entrance to the hive . The studies suggest that in this age can be split in two categories with one continuous variability between them ; the young women they seem to spend more time in the construction of honeycombs and the general maintenance of the colonies , while older bees can go on to process nectar and other tasks , what approaches the entrance to nest [4-6] . The workers close to the entrance are the pillar of the hive 's defense well with your food stored and breeding cannot be relegated to a paper minor in the organization , since the loss of the nest or its content has an impact important on survival and potential colony reproduction [13] .

The assignment of tasks is done through a process synchronized location and broadcast , which allows them keep track of the changes in the demand for labor in the entire colony without the need to communicate with each other about those changes . The activities of these workers they must be tightly related to those of the collectors so that the colony collects 20 kilograms of honey necessary to survive he winter . This is because , although the foragers They collect nectar, transfer it to the workers close to the input to process and store it , and therefore , the number of bees involved when receiving nectar you must be adjusted to match the current feeding rate .

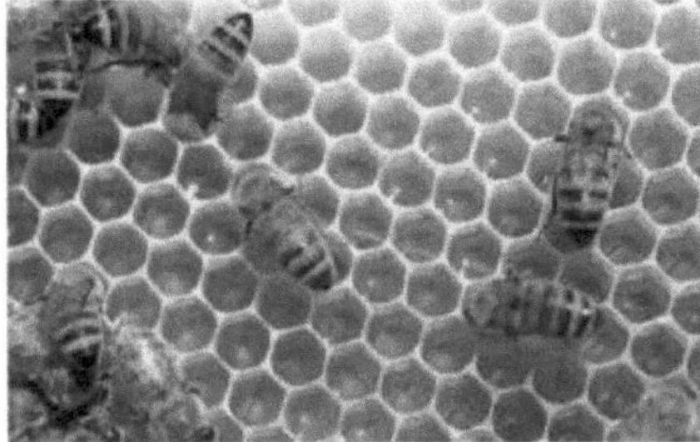

worker bees in honeycomb unloading their crops of nectar (Photography Jose Luis Reyes Carrillo)

Nectar receivers and **honeycomb builders** . The dance or dance of communication , produced for the foragers when determine that there is very few nectar recipients , it serves to recruit more workers to receive it ; are bees also they must build a new honeycomb speed enough to ensure that there is enough space available for incoming nectar [4] ; since the wax is an exudate from four pairs of glands that the workers possess in his abdomen.

workers building wax honeycomb where you can see the whiteness of the cells with fresh wax exuded (Photography Jose Luis Reyes Carrillo)

Guardian bees . A large number of bees prepared to be guardians would give the appearance of inactivity in absence of a colony disturbance. In fact , these workers they can be playing a role very important as booking defensive that can mobilize immediately in he moment of a threat . Nest entrance protection and responses massive flight and sting evolved in response to different types of attacks on colonies. Entry protection is mainly a response to pillage - robbery between hives -, but It is also effective against others invertebrates , such as wasps and bumblebees . The answer mass of the guardians evolved as result of vertebrate predation as the skunks , raccoons , bears and humans [13] .

Lizard invasive dead for the bees guardians and covered with propolis to prevent his decomposition (Photography Hector Genaro Galindo Rodriguez)

In colonies strongly defensive , a small number of workers guardians flies and stings observers humans and animals in the vicinity of the colony; in this behavior can establish a recruitment link , through alarm odors that release the sting detached when biting , in bees close to the interior and those bet on the exit they perceive the smell [14] . It is attributed mainly to isopentyl acetate and 24 other substances found in the poison gland to the alarm reaction and attack of the hive [15] . in bees Africanized was discovered also he 3-methyl-2-buten-1-yl acetate as a new alarm component in the sting [16] .

The need for that complexity chemistry of the alarm signal is not very obvious well some behavioral attributes specifics have been attributed to components individual , but no explanation clear in the set of compounds [17] . This model can be useful to understand bees highly " Africanized " defenses that are much more prone to mass bite events [18] . The differences between the behavior bee defense

European and Africanized could explain by a series of hypotheses . First, the defense in bee colonies africanized can organize the same way as in the European colonies , but the Africanized ones They may be more numerous or more inclined to bite . In second place , the groups of workers other than guardians , can be recruited more easily in answers defensive and third place the bee keepers africanized can be more effective in recruiting bees defenders who counterparty European .

Finally , the defense in bee colonies africanized can organize in a way completely different , with all or most of the colony's workers participating in defense as necessary in instead of having a group differentiated from defenders . The identification of defenders as a group different from workers can help solve one of the main mysteries of the division of labor of bees meKferas , the " bee " phenomenon lazy "; which consists in which many bees pass big proportions of their lives in apparent inactivity . A large number of bees prepared to be defenders would give the appearance of inactivity in absence of a disturbance of the colony, in fact , these bees they would be playing a role very important as booking defensive that can mobilize immediately in he moment of need [13] .

The bees workers guardians they reach this functionality category in the colony around 19 days of age adult , and prior to foraging flights to the fields for transportation of the substances required in the hive .

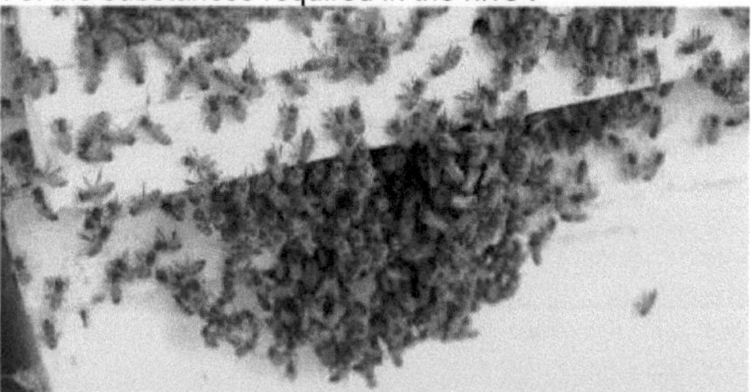

Clustered bees at the entrance to the hive before a threat (Photography Jose Luis Reyes Carrillo)

Foraging bees . Once the job transition is made inside the colony to the forager , the bees of the hive they start to forage in average about 20 days after hatching [5] and these bees they no longer participate in tasks inside the nest ; in change , they focus in look for the four resources that colonies need : propolis , water , pollen and nectar and of the four, the pollen and nectar constitute most of the feeding activity , except in periods of stress by heat in those who collect water can be very important .

Although they have found deviations in the tendency to search a substance specific , almost all bees specialize in pollen or nectar on a given trip in he course of his foraging flight , most bees they seem to be generalists [4] , although some specialize in the collection of plant propolis resinous that in our region can be bushes or trees as the pinabetes , _the governor of ra, the huizac he and the mesqui te , among others .

workers exchanging food , showing pollen load in the corbiculae in the bee at the top , (left image) (Photography Jose Luis Reyes Carrillo) and another collecting propolis in the leaves of the pine tree , (center) and bees propolis pillagers (right) (Photographs Yasmin del Rocio Guevara Ramirez)

Useful life. Bee mortality rates workers differ according to the division of labor , which influences in his useful life . In general, bee mortality in the hive it is low . In contrast to this , mortality rates for foragers have been shown to be substantially higher . The predation , the conditions weather and the possibility of getting lost represent risks extras for the bees that venture outside the hive . The foraging stage of bees meliferas lasts approximately a week and it has been estimated that the mortality rate of these bees beats 15 per hundred diary . Due to the high mortality rate of the workers , to the fact that the time and energy dedicated to foraging is relatively constant among bees workers , the life expectancy of these bees this determined by the age of start of foraging [5] and generally occurs at 20 days of age adult It is not weird Well, the life of a worker , since his hatching of the honeycomb cell to its death , be it between 35 and 40 days .

Communication and foraging . Each anus , a bee colony consumes a considerable amount of food , approximately 20 kilograms of pollen and 60 kilograms of honey . For it They visit and collect - forage - a large number of plants . The thousands of workers foragers in a colony collect laboriously this food from plant communities in flowering that occurs in the field around the hive . Typically, bees in a colony will visit each d^a a dozen or more sources food potentials , each one with his own provision capacity level , determined by variables such as distance from the hive , the abundance and quality of the food offered . To collect his food efficiently , a colony must spread your bees workers among the flowers according to the rewards they offer [19] .

Bee arriving at the cardenche flower (Photography Juan Cabrera Reyes)
communication dance

Reward . The bees foragers returning from a source of nectar or pollen abundant have a greater probability of performing the foraging dance , doing it more forcefully , and returning to the food source more frequently than bees that visit a Poorest source of reward . The colony- level decision to recruit bees that visit a superior quality source or abandon a inferior source , therefore , is a property emergent regulated at the individual level of the forager bee [19] . The workers in waiting to go out foraging they do not attend multiple " dances " that are performed in he frame in an effort by find the one that corresponds to the best food source , but they try only one before leaving the hive in search for the advertised food [20] .

It is known from beginning of the century past the bees meKferas they can learn the time of day in which the flowers secrete the nectar The foragers they return to a food source at the same time in days consecutive and this ability persists during several days after the food source is removed . This apparent memory of time has been tested experimentally and shows that bees can be trained to collect nectar and pollen practically in any time of day The bees synchronize his rhythmic behavior floral daily , feeding only when the nectar and the pollen are at its most abundant levels in the flower In others Sometimes , they remain in the hive , preserving energy that would otherwise be used up in non - productive search flights . The bees in repose they can even park in places relatively distant in the interior of the hive , away from the intense communication dance activity [21] .

Dance types . From he job pioneer of von Frisch and who did it worthy of the Nobel Prize in Physiology and Medicine In 1973, the feeding behavior of bee colonies meHferas could study performing the bee dances individual . The dance or tail dance and the circular dance contain information about where the bees fed dancers and act as bees meHfera he only animal that tells directly to the observer where collected his food . Because only the bees foragers who work with top flowering places perform the dance, not all plant communities in bloom in which the colony is pecoreando are represented by this dance [22] . The bees meKferas They regularly feed several kilometers from the nest , the most common distance is 600 to 800 meters from the hive , the The average is between 2 to 3 kilometers and the circle that surrounds 95 percent Percent of the colony 's feeding activity has a radius of 6 kilometers [23] .

In a study under conditions extremes of scarcity , interpreting foraging distances through the tail dance in the flowering of heather , it was used a hive isolated , located in a forest and it was found that the average search distance was 6.1 km and the distance foraging average It was 5.5 kilometers . Only 10 % of the bees fed within a radius of 500 meters from the hive , while 50 % exceeded 6 kilometers , 25 % more than 7.5 kilometers and 10 % exceeded the 9.5 kilometers away from hive [24] .

This work revealed collection distances extraordinary only explainable due to food shortage in the proximity of the hive and is very It is possible that under arid conditions he collecting behavior for bees is similar.

Bee flying to the flower of the majagua tree (Photography Mario Ruiz Caballero)

A more recent analysis reveals two possible reasons for which the majority of observers Previous participants had not realized the fundamental similarity of the tail dance and the circle dance. First, although the dances that announce food sources far from the hive , that is, up to 500 meters away , they have a very regular pattern of turns. alternate To the left and to the right , the regularity of this pattern decreases in a non - linear mode as the distance between the hive and the food source decreases . It is possible that the observations previous will pass ignore nature unitary dance language because the alternating pattern the turns to left and right is a lot less evident for food sources near the hive .

In second place , the large amount of slight " noise " - in the form of humming in the dances for food sources close - obviously led to an initial conclusion that there was no directional information inside the dance round , as described by von Frisch [22] . However, the current examination of this signaling behavior shows that both the distance information How are the directions? encoded in dances for all distances . However, the signal- to- noise ratio increases as it increases he space and therefore it is possible conclude that bees They have only one dance that they always encodes the distance and direction of the food source , but the precision of the expression of this information It depends on the separation to the recruitment objective [4] .

Water collection . Often the need water of a colony is satisfied with the water that its collectors they recover in passing while collect nectar , since nectar consists mostly in water . Sometimes , however, a part of the colony foragers must gather water intentionally from streams, ponds and other places humid ; since the bees they carry big amounts of water to cool the hive increasing his collection when the highs ambient temperatures make it required cooling of the interior of the colony [25] . It has been found that there are bees in apparent inactivity that are warehouses alive with water , waiting get in in action when the temperature of the hive increase abruptly and the water carriers cannot supply in the need to cool the hive .

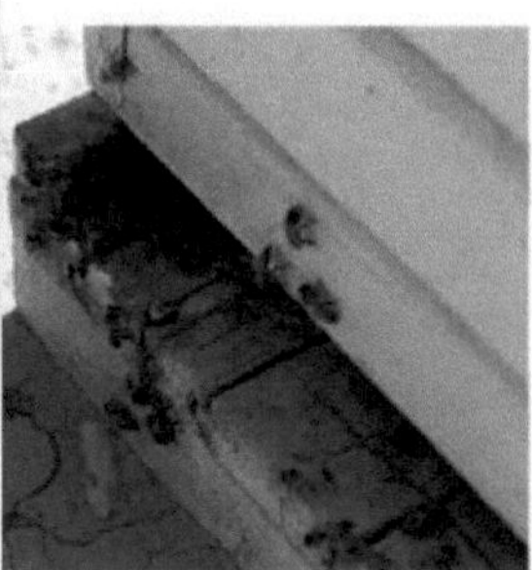

bees collecting water (left image) and workers perched in the piquera ventilating the hive to cool it (right image) (Photographs Jose Luis Reyes Carrillo)
Due to the above, the need for water collection by a hive is therefore very variable , and it is not surprising that a Bee hive possess careful mechanisms to control the speed at which their foragers they collect water . Understand sayings control mechanisms is essential to understand like insect colonies social can respond in a way adaptive to an unpredictable environment [26] .

Workers outside water en an irrigation valve (Photo by Romualdo Basilio Montiel)

The zanganos
Number of drones . bee colonies meHferas are composite by tens of thousands of workers sterile , a single queen and hundreds or thousands of drones [27] . The bee drones meKfera do not forage neither participate in he maintenance or defense of the colony and its only function known is the fertilization of queens vkgenes , which mate with a variable number of drones - which was known to be between 6 and 17 drones during a wedding flight [28] , [29] , but jobs recent They found that it can dock with 34 and up to 77 drones [30] . However, each drone can only do it carry out a time well He dies shortly after mating when his genitals detach from his body , since they act like a plug to give time to which the sperm migrated to the queen's spermatheca [31] where they will be preserved by all his life reproductive .

Drone dead showing part of the intestines when losing the genitals in he flight mating wedding with the queen (Photography Jose Luis Reyes Carrillo)

Origin of the drones Drones are produced from unfertilized eggs that queens they put in the drone cells - a third larger than the worker cell - or laid by the workers , which rarely they lay eggs but can occur when there is no queen ; are ovoposit in the worker cells and give rise to drones small the size of a worker The number of drones found in a colony correlates positively with the number of workers present . The largest colonies they can have up to 1,500 drones adults and their number this limited to a large degree by the amount of drone babies that is produced according to the abundance of food ; However, the colonies will accept big numbers of drones aliens that derive from hives neighbors or that are introduced to the colonies and are accepted for the workers guardians under normal conditions [29] and are more cared for and pampered when the hive this orphan [32] .

The drones request food to the workers and consume honey from the cells , although the workers the feed almost exclusively during the first three days of your life adult , the drones over seven days tend to feed by themselves. They have a feeding rhythm daytime with consumption maximum food before flight activity begins . The flight of the drones It starts between 11:00 and 14:00 and ends between 16:00 and 18:00. Although the maximum flight activity occurs between 2:00 p.m. and 4:00 p.m. , drones even they can start flying from 9:00 in the morning and will join the swarms that leave the colony between 9:00 and 13:00. The time of day in which they fly the drones can be seen determined for the conditions environmental conditions , time of year and possibly for the orientation that they have the beehives of the hives , earlier in which they receive the sun at dawn . If the conditions climatic prevent them from flying the drones

one day , the following flight It will happen earlier the next day [29] .
Useful life. In the drones have reported average lifespans of 13 to 14 days , 21 to 24 days and 34 to 54 days of life . This useful life can vary seasonally in which there are born and can be affected by flight activity or geographic region . The average lifespan of drones tends to be shorter in summer that in autumn and can vary from 13 days in summer in 38 days in fall [29] though the predators they can shorten the life of drones [33] .

Drone about pollen and honeycomb and workers working at your around (Photography Jose Luis Reyes Carrillo)

The survival rates of drones decrease abruptly with him onset of flight activity and probability of survival decreases as it increases he number of flights . The bees workers regulate he number of drones Adults in a colony evicting the drones in autumn and during the periods in which nectar is scarce . The workers They tend to force drones to the honeycombs exteriors of the colony, then to the walls and finally to the lower parts before expelling them from the colony. Certain workers specialize in acts aggressive against drones , the they bite and abuse and some times the They take them out of the hive . The expulsion of the drones is a gradual process that takes several weeks during he autumn . Typically, no more than 10 to 15 drones are evicted from a colony in a single day [29] . In the Lagunera region it is common that after the first autumn or winter frost the entire drones , including Adults young , pupae and larvae , whether Immediately expelled from the hive so that they die of hunger and do not consume the food reserves .

Repose wintry

The behavior of individual bees in the colony changes dramatically as the colony progresses through the different seasons . In late spring, the summer and early autumn , the workers are of a short life - about 30 days - and exhibit a division of labor based in age . The youngest bees , generally less than ten days of age , perform nursing tasks , middle bees age , between 10 and 20 days , they are dedicated to tasks like the construction of honeycombs , the food storage , guarding and whatever is offered , while the older bees in the colony serve in the outside like foragers . In it autumn , as brood decreases , " winter bees " are produced - long life , up to eight months - that will survive the winter . These " winter bees " form he cluster thermoregulator when temperatures drop . Once the breeding of new bees restarts in late winter or early spring , the division of labor resumes among the bees workers that survived the winter [3] .

Enjambrazon

A notable example of coordination and communication is the swarm . reproductive , a process of dividing a colony into which the queen and approximately two thirds or up to three quarters of their bees leave his original nest or maternal colony to form a new colony. As they leave , the workers of a swarm They form a temporary group and start a search for a new nesting site while They are homeless . The workers who remain in the hive they raise one of the old woman 's daughters queen as his new queen [6] .

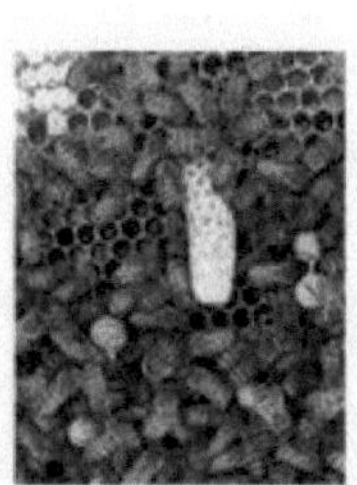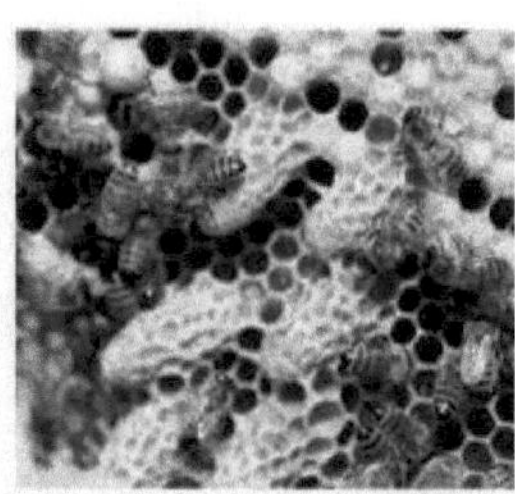

Cells real that they will give origin to a queen next to drone cells , -with operculum bulging - and workers taking care of them (Photographs Yasmin del Rocio Guevara Ramirez)

The development and accumulation of offspring in the spring usually leads to a swarm , where the majority of the workers They leave the colony with the old woman queen in search for a new nesting site , leaving back a new queen and the bees workers remaining to rebuild the original colony. After the swarm , both the original colonies like the new ones pass the rest of the summer and early fall collecting pollen , which is used as source of protein for breeding and nectar , which becomes in honey and used as general energy source , especially during the winter months . When the temperature falls off by below 10 degrees cenrigated , bees in the colony they form a group thermoregulatory ; They form a cluster and vibrate their flight muscles to generate heat that maintains a outer edge temperature greater than 6 degrees cenrigated but usually 12 degrees centigrade . This ensures that the bees in the outermost edges of the group do not cool by under your viable temperature . When breeding begins in winter , the cluster surrounds the breeding area and maintains the core temperature at 33 degrees centigrade . This thermoregulation is achieved only when he cluster this in a space confined , like in he case of natural or made hives by the man [3] .

Conclusions

The queen is a member important of the bee colony meKferas and can be a determinant important health and productivity of the colony . The hives They have a caste system , which involves different types of bees , performing different tasks in the interior of the hive according to age , such as he care of the brood , feeding of the queen and the maintenance of the interior of the hive such as cleaning , thermoregulation and defense . Outside , the bees foragers are in charge of collecting essential substances such as he pollen , propolis , nectar and water necessary for your feeding and cooling . The collectors are capable of communicating to others , through a circular dance and a tail dance , the location , distance and abundance of the source food . The drone or male bee only has a function reproductive , dies when mating with the queen virgin , it is tolerated while exist resources food in the colony and is released out of the hive upon arrival he winter or the season of scarcity . The development of the colony in the spring and acquiring strength with birth of new babies , generally leads to the formation of a swarm , where a large part of the workers They leave the colony with the old woman queen in search for a new nesting site , leaving in the original hive to a new queen and the workers remaining They will rebuild the original hive .

Enjambres de abejas en diferentes arboles (Fotografias Jose Luis Reyes Carrillo)

Literatura Citada

1.	Russell S, Barron AB, Harris D. Dynamic modelling of honeybee (*Apis mellifera*) colony growth and failure. Ecol Modell. 2013;265:158-69.
2.	Seeley TD. Adaptive significance of the age polyethism schedule in honeybee colonies. Behav Ecol Sociobiol. 1982;11(4):287-93.
3.	Doke MA, Frazier M, Grozinger CM. Overwintering honeybees: biology and management. Curr Opin Insect Sci. 2015;10:185-93.
4.	Johnson, BR. Division of labor in honeybees: form, function, and proximate mechanisms. Behav Ecol Sociobiol. 2010;64:305-16.
5.	Knoll S, Pinna W, Varcasia A, Scala A, Cappai MG. The honeybee (*Apis mellifera* L., 1758) and the seasonal adaptation of productions. Highlights on summer tow in tertransition and back to summer metabolic activity. Livest Sci. 2020;235:104011.
6.	Grozinger CM, Richards J, Mattila HR. From molecules to societies: mechanisms regulating swarming behavior in honeybees (*Apis* spp.). Apidologie. 2014;45:327-46.
7.	Slessor KN, Winston ML, Le Conte Y. Pheromone communication in the honeybee (*Apis mellifera* L.). J Chem Ecol. 2005;31(11):2731-45.
8.	Kocher SD, Grozinger CM. Cooperation, conflict, and the evolution of queen pheromones. Review. J Chem Ecol. 2011:37:1263-75.
9.	Slessor KN, Kaminski LA, King GGS, Borden JH, Winston ML. Semiochemical basis of the retinue response to queen honeybees. Nature. 1998;332(6162):354-6.
10.	Kocher SD, Ayroles JF, Stone EA, Grozinger CM. Individual variation in pheromone response correlates with reproductive traits and brain gene expression in worke rhoneybees. PloSOne. 2010;5(2):e9116.
11.	Walton A, Toth AL. Variation in individual worker honey bee behavior shows hallmarks of personality. Behav Ecol Sociobiol. 2016;70:999-1010.
12.	Walton A, Dolezal AG, Bakken MA, Toth AL. Hungry for the queen: honey bee nutritional environment affects worker pheromone response in a life-stage dependent manner. Funct Ecol. 2018;32(12):2699- 706.
13.	Breed MD, Robinson GE, Page RE. Division of labor during honey bee colony defense. Behav Ecol Sociobiol. 1990;27:395-401.
14.	Breed MD, Guzman-Novoa E, Hunt GJ. Defensive behavior of honeybees: organization, genetics, and comparisons with other bees. Ann Rev Entomol. 2004;49:271-98.
15.	Free JB. Pheromones of social bees. Chapman and Hall, London, UK. 1987;236p.
16.	Hunt GJ, Wood KV, Guzman-Novoa E, Lee HD, Rothwell AP, Bonham CC. Discovery of 3-methyl-2-buten-1-yl acetate, a new alarm component in the sting apparatus of Africanized honey bees. J Chem Ecol. 2003 Feb;29(2):453-63.
17.	Trhlin, M, Rajchard J. (2011). Chemical communication in the honeybee (*Apis mellifera* L.): a review. Vet Med. 2011;56(6):265-73.
18.	Guzman-Novoa E, Hunt GJ, Page RE, Uribe-Rubio JL, Prieto-Merlos D, Becerra-Guzman F. Paternal effects on the defensive behavior of honeybees. J Hered. 2005;96:376-80.
19.	Seeley TD, Camazine S, Sneyd J. Collective decision-making in honeybees: how colonies choose among nectar sources. Behav Ecol Sociobiol. 1991;28:277-90.
20.	Seeley TD, Towne WF. Tactics of dance choice in honeybees: do foragers compare dances? Behav Ecol Sociobiol. 1992;30:59-69.
21.	Moore D. Honey bee circadian clocks: behavioral control from individual workers to whole-colony rhythms. J Insect Physiol. 2001;47:843-57.
22.	von FrischK. The dance language and orientation of bees. Harvard University Press, Cambridge, MA.1967;566p.
23.	Visscher PK, Seeley TD. Foraging strategy of honey bee colonies in a temperate deciduous forest.

Ecology. 1982;63:1790-801.
24.   Beekman M, Ratnieks FLW. Long-range foraging by the honey-bee, *Apis mellifera* L. Funct Ecol. 2000;14:490-96.
25.   Kuhnholz S, Seeley TD. The control of water collection in honey bee colonies. Behav Ecol Sociobiol. 1997;41:407-22.
26.   Gordon DM. The organization of work in social insect colonies. Nature. 1996;380:121-24.
27.   Brutscher LM, Baer B, Nino EL. Putative drone copulation factors regulating honeybee (*Apis mellifera*) queen reproduction and health: a review. Insects. 2019;10:1-18
28.   Woyke J. Multiple mating of the honeybee queen (*Apis mellifica* L.) in one nuptial flight. Bull AcadPolonSci Cl. 1955;3:175-80.
29.   Currie RW. The biology and behaviour of drones. Bee World. 1987;68(3):129-43.
30.   Withrow JM, Tarpy DR. "Cryptic royal" subfamilies in honeybee (*Apis mellifera*) colonies. PLoS ONE 2018;13:e019912.
31.   Woyke J, Ruttner F. Ananatomical study of the mating process in the honey bee. Bee World. 1958;39:3- 18.
32.   Langowska A, Zduniak P. No direct contact needed for drones to shorten workers lifespan in honey bee. J Apic Res. 2019;59:88-94.
33.   Rueppell O, FondrkMK, Page RE. Biodemographic analysis of male honey bee mortality. AgingCell. 2005;4:13-9.

And when he autumn I arrive , and I arrive also he ended his days , he had still

time to give one last lesson before dying to the young girls bees that

surrounded him : It is not ours intelligence , but our job who us makes so strong

Horacio Quiroga

3. Number of bees entering the hive and capturing pollen

Jose Luis Reyes-Carrillo, Jose Luis Galarza Mendoza and Juan Cabrera Reyes

Introduction

The bees meKferas are attracted by the stimuli visual and olfactory characteristics of flowers [1] . The compounds aromatics become progressively easier for bees to identify with the increase in the concentration of the odor . This makes it easier also the discrimination between some aromas and others [2] . The pollinators they visit sequentially and faithfully the flowers of a same species still when fly over or there near other flowers species in reward function . This " floral constancy " of foraging behavior has been described mostly in meKferas bees [3] given that their activity in the field can be determined very various forms [4] .

In a hive , at the same time , different groups of bees are dedicated to tasks you specify as he care of the offspring , the construction of combs , the defense of the colony, the maintenance of the temperature of the brood nest as well as collect and process nectar and pollen . Because in each moment there are hundreds of individuals developing different labors Simultaneously , some of those chores They must be coordinated for the colony to survive [5] . Asl the bees adjust his pollen foraging activity according to the needs of the colony , which are certain by the amount of pollen stored and by the amount of breeding young since he pollen collected is deposited in the cells watts of the backstage in those with the greatest amount of breeding open , regardless of the place of this frame inside the hive . The bees also they inspect a greater number of cells in that rack and spend most of their time there , which is a evidence that pollen foragers They have the ability to evaluate the pollen requirements of colony [6] and therefore to ensure that the supply is not lacking .

Coordination of activities in a hive It happens in two ways : first there is the coordination between those participating bees in a same work, and in second , that of those that carry out labors different but complementary than some manner they must establish communication links [7] that allow sync up his work [8] . For example , workers that collect nectar must coordinate their activities with nurse bees [9,10] . Studies recent about the food supply processes , construction behavior and social differentiation in insect societies or societies of others arthropods gregarious , show the importance of self-organization mechanisms [5,10,11] . One measure of the strength of a colony is the population. existing and therefore it will be observed a greater number of bees leaving and entering through the piquera when the hive It is more populated . Even when it is known he behavior productive of the hive , in the Comarca Lagunera it was not known how much pollen collect the bees in the autumn season and yes he number of bees entering through the piquera , like population indicator , kept some relationship with the amount of pollen harvested . Are questions were resolved studying he pollen entry into the hive through field observations .

Study

The present work was carried out in he apiary Technological Institute school Agricultural No. 10, currently Torreon Technological Institute , in the Anna ejido, municipality of Torreon, Coahuila, from October 13 to 26 to determine the relationship between the number of bees entering by minute , which is an indicator of the amount of population of the colony, and the amount of pollen collection during Fall [12] . They were used eight bee hives Jumbo size , each a equipped with a pollen trap modified Ontario type [13] . At the beginning he experiment , first it was determined he number of bees entering the hives by minute at twelve o'clock by direct observation in the corner of each hive .

Bees entering the hive through the piquera before placing pollen traps (Photography Jose Luis Reyes Carrillo)

Determining the number of bees entering by minute it was done five times consecutive and obtained he average , to stay as go on :
Number of bees entering the hive by minute before placing pollen traps Modified Ontario type .

hive	bees
1	163
2	143
3	114
4	159
5	133
6	107
7	98
8	97
addition	1014
average	126.7

I agree with you data collected , the minor number of bees entering It was 97 and the maximum was 163 bees. by minute . The overall average of the eight hives was 126.7 bees by minute .
The pollen was harvested individually and daily in each hive at sunset , and separated by sizes according to his diameter through the meshes laboratory standard of 2.04 miKmeters (pollen large), 1.91 miKmeters (pollen medium) and 1.19 miKmeters (pollen small) and then heavy fresh in a digital laboratory scale . An analysis was done statistics of the results [14] .
Results

In the foreground a hive about a pollen trap modified Ontario type in which the entry of the bees is observed . In the background hive you can distinguish the tray where it accumulates he pollen (photography Juan Cabrera Reyes)

Pollen amount harvested

The amount of pollen harvested during he sampling month I present ups and downs very marked with a decline particularly accentuated in the last week . The severity of the changes prevented find a consistent capture pattern in he time , since they are observed ups and downs in pollen collection in the next graph :

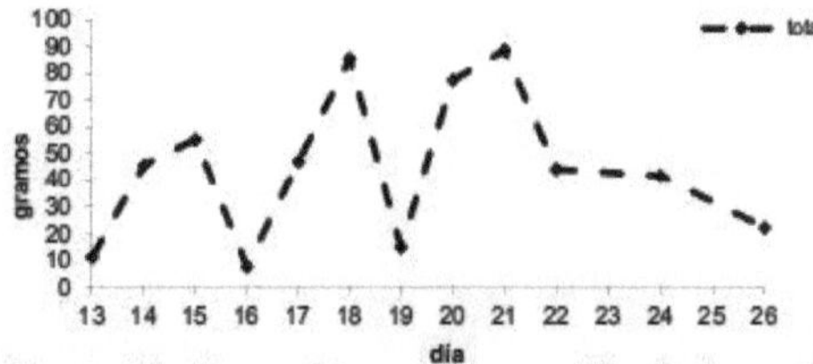

day

Average total corbicular pollen by eight o'clock day hives with traps guy *Ontario modified for month of October*

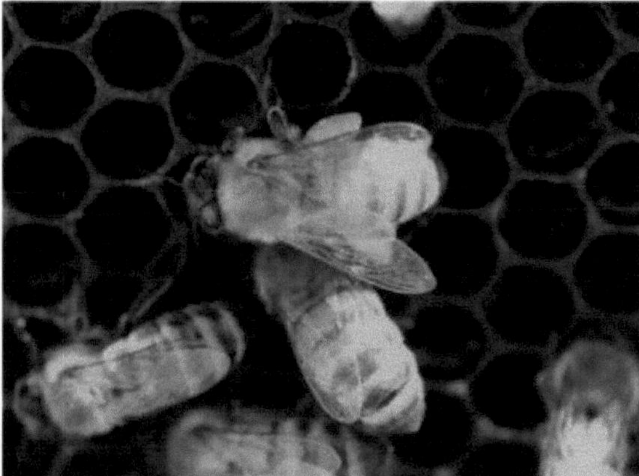

Bee with light yellow pollen on the corbuculae (Photography Jose Luis Reyes Carrillo)

Between pollen separate by sizes only found cumulus medium (1.91 mm) and

43

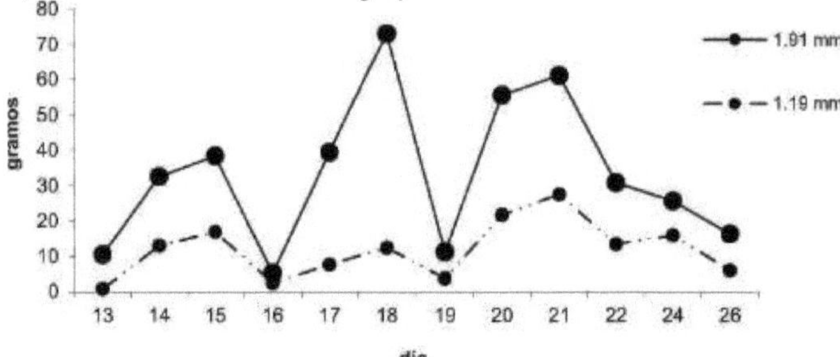

*corbicular pollen medium (1.91 mm) and small (1.19 mm) in hives with traps
modified Ontario type in he month of October*
*During the time of the study , it was possible observe that the ups and downs in
the amount pollen harvest showed a two- week pattern of increase in the amount
of pollen and a week of decrease very marked . This was repeated by three times
until the last week , when it ended he observation period .*

Modified Ontario Pollen Trap Tray showing he pollen recovered from the corbaculae of
bees (Photography Juan Cabrera Reyes)
Since the main objective of the work was relate he number of bees entering the
hive with the amount of pollen collected , observations can be represent
graphically as go on :

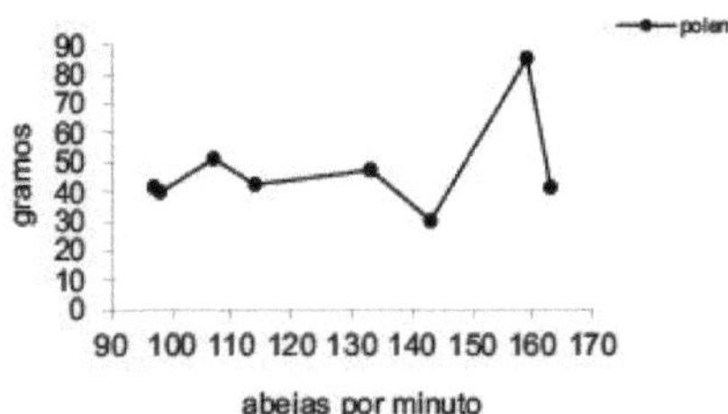

Pollen amount average diary in relation to the number of bees entering by minute
to the hive in he month of October
It was not observed a trend defined pollen capture , since the relationship
between the number of bees entering by minute to the hive with the grams of
pollen accumulations was not very defined and therefore without any predictive
value ; that is, it could not be determine the amount of pollen to be harvested per
colony counting he number of bees in the piquera ; this can because when you
had a colony with 159 bees by minute pollen collection It was the maximum, but

decreased to levels similar to less hives entry of bees and it was same as with him maximum number of bees entering , which was 163. This showed a inconsistency that I did not allow associate are quantities with the number of bees .

Regarding the quantity pollen average collected by day by hive during he month of October , the comparison of the averages indicate a difference only for the value of 143 bees by minute , which was the oldest. For the rest of the colonies, the pollen harvest was equal .

Number of bees entering the hive by minute and quantity pollen average by d^a during he month of October

hive	bees	average (g)
1	163	41.4
2	143	85.3
3	114	30.1
4	159	47.3
5	133	42.3
6	107	51.0
7	98	39.9
8	97	41.5
addition	1014	378.8
average	126.7	47.4

This behaviour erratic can due to time dedicated to attend he request colony water . Since the workers they collect a water amount important to cool the hive , they increase his collection when the temperatures daytime are high well require cool the hive and decrease his haulage when happens he period critical heating [25] . This could be an explanation for the ups and downs in pollen collection , since bees they would be busy carrying water and suspend he carrying of others hive products , such as he pollen , nectar and propolis . It is possible that during spring and summer , seasons in which bees reproduce in greater quantity and demand a greater volume of food , their inbound traffic activity reflect better his strength .

Conclusions

During the month of October the amount of pollen collected by hive did not show a consistent pattern and decreased at the end of the month . The bees collected a greater amount of pollen clusters of the same size medium . The number of clusters of sizes medium and small increased and decreased similarly . The comparison between the amount of pollen harvested by hive and the number of bees observed entering the hive by I don't save a minute relationship some .

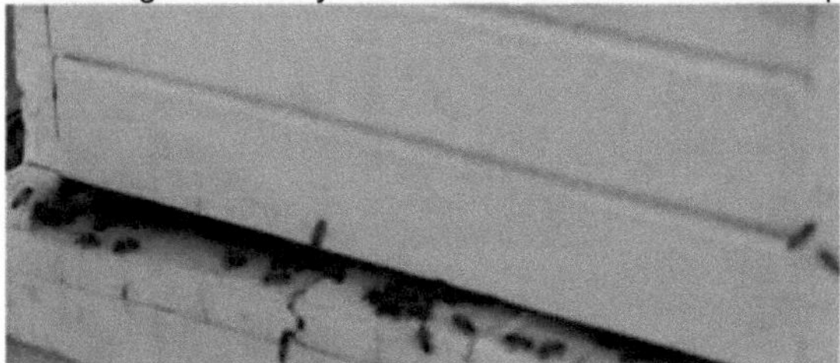

Bees going in and out in the hive (Photo by Jose Luis Reyes Carrillo)

References

1.	Galizia CG, Kunze J, Gumbert A, Borg-Karlson AK, Sachse S, Markl C, et al. Relationship of visual and olfactory signal parameters in a food-deceptive flower mimicry system. Behav Ecol 2005;16: 159-6

2.	Wright GA, Smith BH. Variation in complex olfactory stimuli and its influence on odor recognition. Proc R Soc London B 2004;271:147-52 .

3.	Gegear RJ, Laverty TM. The effect of variation among floral traits on the flower constancy of pollinators. In: Chittka L, Thomson JD. Editors. Cognitive ecology of pollination. Animal behavior and floral evolution. Cambridge University Press. Cambridge, U.K. 2001;1-20.

4.   Hagler JR, Jackson CG. Methods for marking insects: Current techniques and future prospects. Annu Rev Entomol. 2001;46:511-43.

5.   Amdam GV, Norberg K, Fondrk MK, Page RE. Reproductive ground plan may mediate colony-level selection effects on individual foraging behavior in honey bees Proc Nat Acad Sci USA. 2004;101:11350- 55.

6.   Dreller C, Tarpy DR. Perception of the pollen need by foragers in a honey bee colony. Anim Behav 2000;59: 91-6.

7.   Tautz J, Casas J, Sandeman D. Phase reversal of vibratory signals in honeycomb may assist dancing honey bees to attract their audience. J Exp Biol 2001;3737-46.

8.   De Marco RJ, Farina WM. Trophallaxis in forager honeybees (*Apis mellifera*): resource uncertainty enhances begging contacts? J Comp Physiol A Neuroethol Sens Neural Behav Physiol. 2003;189:125-34.

9.   DeGrandi-Hoffman G, Hagler J. How honey bees might use the placement of incoming nectar in a colony as a means of communication. Am Bee J 2000;140:892-4.

10.   Farina WM. The interplay between dancing and trophallactic behavior in the honey bee *Apis mellifera*. J CompPhysiol. 2000;186:239-45.

11.   Farina WM, Wainselboim AJ. Changes in the thoracic temperature of honeybees while receiving nectar from foragers collecting at different reward rates. J Exp Biol. 2001;204:1653-58.

12.   Orozco-Vidal   JA, Reyes-Carrillo JL, Cabrera-Reyes J, Galarza-Mendoza JL. Numero de abejas
entering the   hive and capturing pollen . Report of the 14th International Update Congress
Apfcola , Boca del Rfo , Veracruz. May 16-18, 2007;146-50.

13.   Waller GD. A   modification of the OAC pollen trap. Am Bee J 1980;120:119 -21.

14.   Steel RGD, Torrie JH. Principles and procedures of statistics. McGraw-Hill Book Company, Inc. 1960;New York, USA:481p.

15.   Kuhnholz S, Seeley T. The control of water collection in honey bee colonies. Behav Ecol Sociobiol 1997;41: 407-422 .

" Although honey this in each flower , always does lack a bee to get it out "

Guiterman

4. The bees collect materials strangers to the pollen during the era of scarcity

Jose Luis Reyes-Carrillo, Jose Luis Galarza Mendoza and Juan Cabrera Reyes

Introduction

Pollen identification in a pollination study It serves various purposes . When analyzing he pollen they carry the visitors florals you can find the evidence on the variety of species visited [1,2] . The pollen in stigmata can indicate Yeah he pollen deposited is of the same species [3] , if there is a potential stigma blocking by he pollen from others species or others similar effects [4] . Pollen identification is also important in studies that measure the patterns pollination seasons masculine and feminine . Researchers studying anemophily - pollination by he pollen transported by he wind - can need identify the species present in their samples with the object of knowing the different percentages of species found or determined the pollen contents of species allergenic in the air [5] . The characterization of pollen fossil provides a record of the flora and climate in the past [6] .

The bees consume he pollen from flowers and provide a benefit directly to the plants through their collection activities 7 · The plants that produce seeds are pollinated by animals and generally have pollen large , sculpted and covered with an adhesive wax or some substance oily The cover causes the pollen grains to adhere to each other and to the animals pollinators , feed the herbivores , attract pollinators and be a good source of food for them [8] .

In a pollen collection study in the Comarca Lagunera , when collecting tray samples recipients in a trap modified Ontario type , they could notice materials vegetables that were identified such as parts of the governor plant, dried parts of plants or seeds that look like small bones , seeds of vegetables that were identified like from someone type of grass , plant parts unknown , pods full of grass seed known such as " mostacilla " or " chili de pajaro " and anthers sorghum completes ; some bees they got enter the trays receptors of the pollen trap , leaving trapped in she with the pollen ball fitted in he sting , surely as strategy to prevent the pollen trap from retaining them and this was observed in different dates [9] .

Bees with their respective pollen ball fitted in he sting , seed adhered to propolis particle and governora plant tissue found in the pollen traps (Photographs Jose Luis Reyes Carrillo)

Under arid and low conditions temperatures in he winter of the Comarca Lagunera [10] , in this station is the smallest presence of flowering in floors wild or cultivated and bees so could obtain materials other than pollen in his food collection .

The above is shown in this chapter product of an investigation carried out in this region [11] where the question is answered Yeah in bees are scarce they collect materials alien to pollen

Study

The present work was carried out in he apiary Technological Institute school Agricultural No. 10, currently Torreon Technological Institute , in the Anna ejido, municipality of Torreon, Coahuila during the winter months of January , February and March with the purpose of checking yes the bees they collected substances other than pollen during the time of scarcity . 12 hives were used Jumbo type equipped each one with one pollen trap modified Ontario type [12] harvesting he pollen weekly . The pollen was separate by size according to his diameter through the meshes laboratory standard of 2.04 millimeters (pollen large), 1.91 millimeters (pollen medium) and 1.19 millimeters (pollen small) and then heavy fresh . The substances alien to pollen were separated , identified and weighed in fresh.So much the items alien as he pollen were weighed in a digital laboratory scale . The data of the number , weight and quantities respective were processed statistically [13] .

Cells containing pollen of different colors (Photography Yazmin del Rocfo Guevara Ramfrez)

Results

Items alien to pollen

The elements unrelated to pollen , the fundamental reason for the study , appeared from the first week , revealing between them seeds tiny of various plants , leaflets dried from the mesquite plant , which are common in the region and where they were near the apiary several trees and bushes of this species . Sayings trees for this date They presented the dry leaf due to winter frosts and fall of the leaflets .

were also found pieces small colored sheets light green and reddish of unidentified plants which could not be located in the vicinity of the apiary and that they could correspond to ornamental plants of one 's own school or the Anna ejido , located one kilometer away approximately .

Cones casuarina males were found in the pollen traps and compared to the pollen cones nearest trees - approximately 500 meters away - coinciding with their flowering time which suppose they were removed from the trees by the bees and transferred to the hive , which explains his presence in the pollen trap .

They met and separated stems dried , wood chips , clumps the size of pollen grains , plant material that looks like bones and stingers loose from the bees . Some of them could be part of the waste that bees they wanted remove from the hive and they were thrown or fell accidentally in pollen traps . They can be see in the following Photographs :

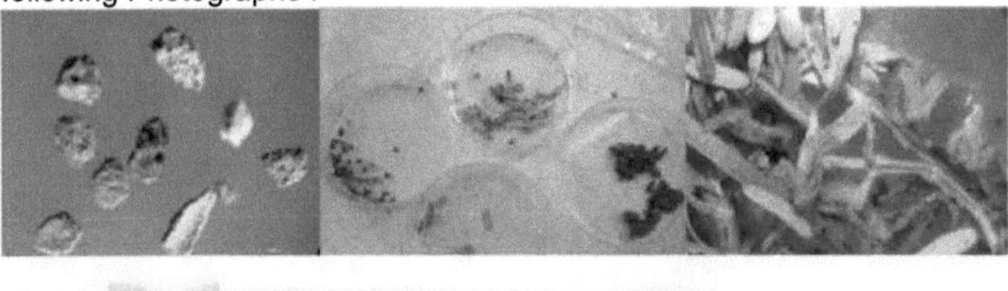

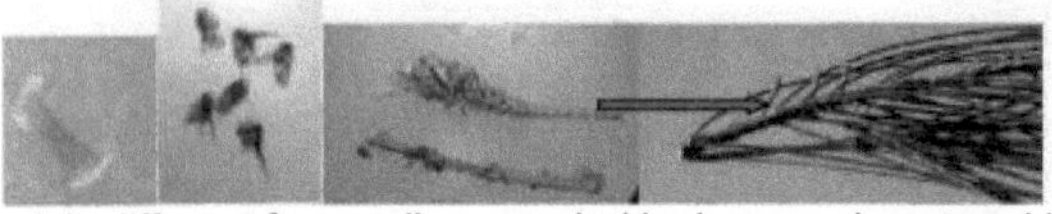

Materials different from pollen , carried by bees and captured in the pollen traps (photography Juan Cabrera Reyes)

In it following picture are shown the items found in pollen traps during he period winter and their respective quantities :
Number of elements alien to pollen captured by week in the 12 pollen traps during the winter months from January to March . Torreón, Coahuila.

week	seeds	leaves/ leaflets pieces /sheet	clods	stems splinters	cones pollen	propolis flakes	stingers
Jan-07	62	81	32	1	2	64	
Jan-13	10	14	17			19	
Jan-22	18	17	2			6	3
Jan-28	4	13		5		17	
Feb-04	2	12		2		4	
01-feb	1	5		2	1	13	1
feb-18	1	83		17		84	
feb-25		64	4	24		48	
mar-04		16		6		31	2
mar-11	1	14		4		4	
mar-17	6	28	7	7	3	69	
total	105	347	62	68	6	359	6
average	11.7	31.5	12.4	7.6	2.0	32.6	2.0

Soil particles, similar in size to pollen belong to the typical floor sodium in the area and with a high salt content , which could explain the reason why bees they collected said material to use it as a complement nutritional

When isolating the materials foreign pollen collected , they separated Also the propolis particles , large and small, that fell into the traps and give an idea of the activity. promoter within the colony, which consists in he sealing gaps , gluing hive frames and lid on the inside . Can point out that the propolis observed was a mixture of flakes , particles and pieces clumps that fell from the bees ' legs when they were carried to the hive rear and could not get them back through the mesh that prevents his access to tray collector .

Deprive bees of varied natural food sources can have the unintended consequence of increasing his acceptance of food from another way they would be rejected [14] . It is known that bees not only detect toxins , but also they can learn to associate their smells with their flavors and with the consequences of consuming them [15] . However, compared to other insects the bee It has only 10 receivers gustatory and therefore its sense of taste is very poor compared to the fruit fly which has 68 receptors taste and the anopheles mosquito that has 76 [16] .

It is likely that the items vegetables such as leaves or fragments of them , mesquite leaflets , and small stems , bean could have been torn off and transported intentionally for the same bees in need of food , and not accidentally fell into the trays , although this could not be verified through observations in field .

The above noted , although speculative , is based in careful field observations accompanied by examinations microscopic of others studies that provided evidence convincing that it involved the bee meHfera as the cause of the perforation of the leaf in birch species . These revealed that bees they cut or detached selectively pieces of leaves and shoots youths in search for one substance sticky , apparently propolis .
Later , as they matured The cuts grew larger and caused he tearing of the leaves. The studies biological They indicated that this freak reached its maximum point at the station dry , with an index of up to 90% of young leaves damaged by outbreak by month and was lower in the wet months during he leaf emergence peak . The damage caused to leaves by bees was consistently higher in the birch species in the Little trees youths more than the mature trees and on leaves exposed to the sun more than on shaded leaves [17] .

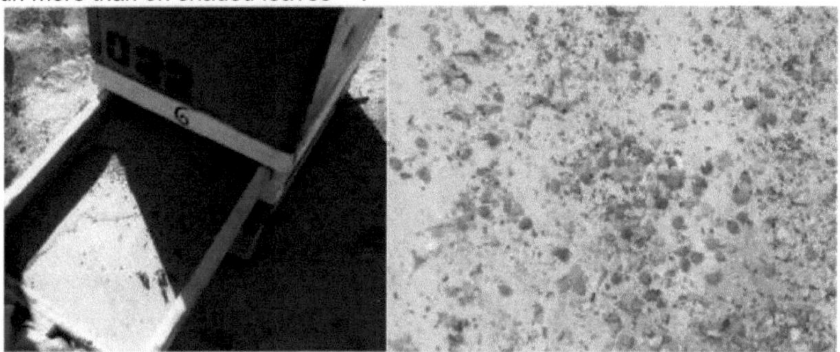
Pollen Trap Tray containing the pollen clusters , items alien to pollen and bees dead (photographs Juan Cabrera Reyes)

Pollen amount
The amount of pollen collected through the hives can consider it low then corresponds to winter . The data presented is weekly for all 12 colonies and is shown in the next board :
Amount of pollen , substances alien harvested and percentage by week in 12 hives provided with a pollen trap during the months of January to March . Torreón, Coahuila.

week	pollen (g)	substances foreign (g)	%	observations
January 7th	152.8	36.60	24.0	
January 13th	16.1	11.36	70.4	rain
January 22	302.0	27.28	9.0	
January 28	64.1	16.28	25.4	rain
February 4	95.6	26.79	28.0	rain
February 11th	325.7	29.82	9.2	
18th of February	423.0	40.22	9.5	
February 25	338.8	63.31	18.7	
March 4	282.9	34.24	12.1	
March 11th	221.8	46.88	21.1	
March 17	269.0	30.61	11.4	
total	**2491.7**	**363.39**	**14.6**	

average weekly	**226.5**	**33.0**
average hive	**18.9**	**2.8**

Week in which it was obtained less pollen it was the second week of January , corresponding to the first month of the study , and coinciding with rains light abnormal for this time of year they were presented that week .

During the month of January and beginning of February were recorded temperatures lows and rains light that prevented the normal activity of the bees in pollen foraging in the species vegetables that develop in the region during he period winter . Since since he December temperatures hafran dropped below 0 °C, the remnants of summer vegetation already They have disappeared.

Autumn - winter species common in the region , like mostacilla or grass in cross and mallow [18] that tolerate frost light , appear usually associated with crops perennials like alfalfa that is irrigated all he year , in addition to forage crops winter annuals such as oats , triticale, rye grass and wheat. Are species They could be the source of supply of the pollen that appeared in the traps , but his origin botanist was not certain .

The amount of pollen collected by week of the 12 hives shows that the amount was very variable since the highest amount was 423 grams - second week of February - and the lowest of 16.1 grams , corresponding to the second week of January . However the average by hive It was only 18.9 grams .

The data from the 11 pollen collections reveal that the study period I represent for the bees a epoch severe shortage , given that for the Lagunera region in pollen trap hives during melon pollination , in average the normal harvest per week is 200 grams of pollen accumulations during spring and summer . In the Laguna region in the vegetation wild with beehives away from the crops , were obtained in average quantities of 430 grams by week in spring, 361 grams in he summer , 76 grams in autumn and 85 grams of pollen corbicular in winter [19] .

Previous amounts of pollen are related to the demand for the amount of offspring that the colony has and to which the workers react foragers carrying the necessary proportion [20] . In the different Seasons of the year vary according to the population of the hive , being in autumn and winter when the queen decreases his posture in the face of lack of flowering and losses temperatures , and Therefore , there is a minor need for food protein by having minor breeding quantity each colony.

captured pollen in the traps showing the color variation (Photographs Juan Cabrera Reyes)

When comparing the grams of pollen harvested In total of the 12 hives, it is observed that there was no a relationship that could associate the amount of substances extraneous with respect to the amount of pollen carried to the hive well he increase in grams of these products foreign to pollen do not correspond in manner some with variations in pollen quantities caught in traps , as illustrated in the next graph :

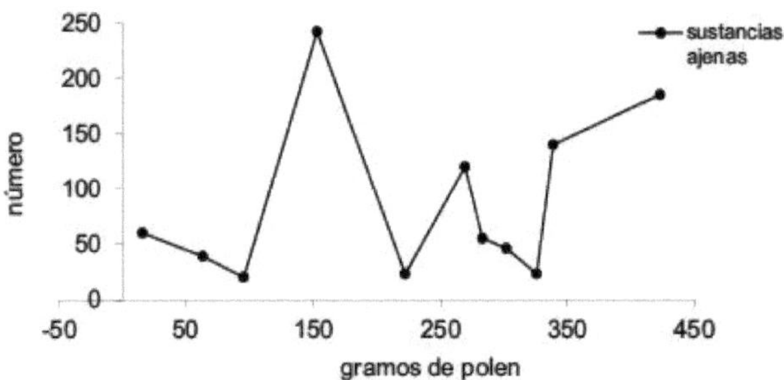

Number of substances alien to pollen in relation to the amount of pollen harvested in the traps of 12 hives

A wide variety of materials in pollen loads corbicular has been registered in the literature . Microscopic examination of bee pollen loads meHferas trapped in the field and pollen collected in traps in the hives they have revealed constituents additional as petals , filaments and stamens completes of a plant, possibly canola, cellulose filaments from non- lignified tissues of the plant called Bugloss they were part of the anthers , and tiny insects like thrips, both adults as larvae [21] . These last When they are larvae they feed on plant tissues , and adults have a diet varied based mostly in he pollen . It is assumed that for his size little these insects were packed accidentally by the bees when making the pollen ball .

They have found stamens complete or anthers in pollen loads corbicularis of many species vegetables particularly of the grasses , plants umbeHferas and plantago and clovers [22] . In hives of the Lagunera region Near the corn and sorghum, the pollen balls are easily observable. corbicular with anthers than bees they carry both crops to their hive .

Size of loads , clusters or pollen pellets

With respect to different sizes of the clusters or pellets of pollen , which was he second approach to this study , it was observed that in the weeks that there was minor catch amount , most corresponded to clusters of size small , - week of January 13 and January 28 -. For the others evaluation dates predominated those of size intermediate , as can be notice in he following chart :

Amount of pollen of different size harvested by week in 12 hives during he winter

week	2.38mm	1.91mm	1.19mm	weekly total
January 7th	55.8	90.1	6.9	152.8
January 13th	0.0	0.0	16.1	16.1
January 22	52.8	135.4	113.8	302.0
January 28	4.3	22.5	37.3	64.1
February 4	11.4	50.9	33.3	95.6
February 11th	87.8	189.4	48.5	325.7
18th of February	78.3	284.9	59.8	423.0
February 25	48.4	219.2	71.2	338.8
March 4	34.3	183.7	64.9	282.9
March 11th	55.9	135.7	30.2	221.8
March 17	123.4	100.7	45.0	269.0
total	552.1	1412.5	527.1	2491.7
average	50.2	128.4	47.9	226.5
percentage	22.2	56.7	21.2	

When it was obtained he average of the three sizes of the accumulations for all harvest weeks were found differences , that is, the size clusters intermediate were the larger ones and the smaller ones big and small were equal between sL

Standard laboratory meshes to separate by size the pollen clusters . They weighed themselves so much pollen like substances unrelated to pollen (Photographs Juan Cabrera Reyes)

When referring in percentage of pollen amounts corbicularis it was observed that the size clusters small -1.19 mm- and those of size large -2.38 mm- were equal in quantity and that the majority corresponded to the size clusters medium -1.91 mm-. In a study to determine he pellet size carried in the bees ' corbuculae the surface area of each pellet was from 1,543 to 2,281 square miKmeters [23] , which coincides with the dimensions obtained , with respect to the smallest and largest clusters that were obtained and measured with the meshes in this study .

On a related idea about impulsivity , which involves elections fast and inaccurate about the slow and precise ones , it has been suggested that the degree of behavior impulsive in bees and others insects social correlates positively with the need nutrition that the animal faces . In bees , a metabolic regulation of impulsivity could mean that the foragers can become more impulsive due to malnutrition resulting from factors ecological such as scarcity and causing them not to take advantage all he floral resource [24] .It is known that to take away pollen to the colony, the workers foragers with larger corbuculae they collect pollen that compresses in larger size pellets and then they can do less trips to the colony, dedicate less time to explore in search for flowers and spend more time on pollen sources already located . However this is not relevant when he pollen is so scarce that a worker ca n't pick up neither a full load of pollen . So he size of the pollen basket on the bee 's leg would not be important [23] but the conditions of the region at the time winter .

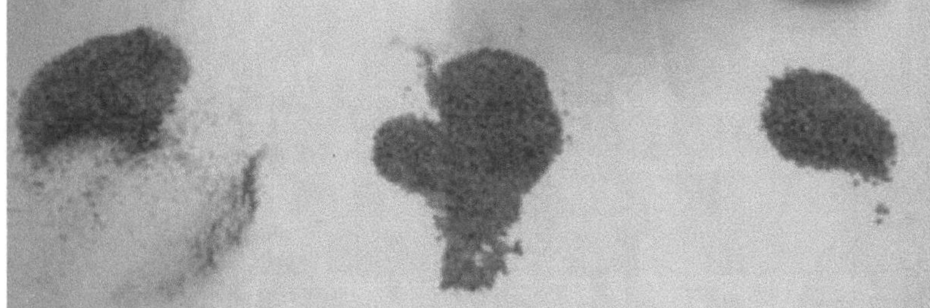

Pollen pellets of different sizes separated in the laboratory meshes , from left to right : large (2.38 mm), medium (1.91 mm) and small (1.19 mm) respectively (Photography Juan Cabrera Reyes)

Weekly pollen total collected was variable, with an average of 226.5 grams , highlighting he month of January due to the shortage , with the 2 lowest weeks of collection in this period winter . This could associated with the presence of rain in those dates , which prevented the workers go out foraging .

It is estimated that a bee colony consumes around 20 kilograms of pollen per year [25] , which would correspond to about 384 grams. by week , although it is understood that depending on the season of the year has to increase this amount to the extent that there are more babies to feed and there is a greater amount of resources , and, decrease when there is little queen posture and scarcity of available flowers .

The efficiency of a pollen trap to capture pollen loads is around 16% . As he flow as he size of loads can vary considerably depending on factors such as the time of day , the plant species visited and the conditions meteorological . Are last they can forcing the foragers to return to the hive before they had filled a load [26] and that could be a good estimator of the total volume harvested by the workers of a hive .

Conclusions

During the winter it was found that the amount of pollen collected was very low. In addition to pollen , bees they collected seeds , leaves, leaf fragments , wood chips , small stems , clods of soil salty and cones male casuarina. The substances alien collected through bees you can attribute to lack of food . These materials oblivious to the pollen they did not keep relationship some with the amount of pollen caught in the traps . The result coincides with observations in others studies in those who bees they collected substances alien to pollen in the time of scarcity . The pollen accumulations corbicular collected for the bees were of dimensions varied , in his majority of size medium , and, in equal size quantities big and small .

Worker bees with loads of pollen corbicular on the legs and cells with pollen already packed (Photography **Yazmin del Rodo Guevara Ramfrez**)

References

1.   Bryant VM, Pendleton M, Murry RE, Lingren PD, Raulston JR. Techniques for studying pollen adhering to nectar-feeding corn earworm (Lepidoptera:Noctuidae) moths using scanning electron microscope. J Econ Entomol. 1991;84:237-40.

2. Goodwillie C. Wind pollination and reproductive assurance in *Linanthus parviflorus* (Polemoniaceae), a self-incompatible annual. Am J Bot. 1999;86:948-54.

3. Takayama S, Shiba H, Iwano M, Shimosato H, Che F, Kai N, et al. The pollen determinant of self - incompatibility in *Brassica campestris* . Proc Natl Acad Sci USA. 2000;97: 1920-2

4. Tandon R, Manohara TN, Nijalingappa BHM, Shivanna KR. Pollination and pollen-pistil interaction in oil palm, *Elaeis guineensis* . Ann Bot. 2001;87: 831-8

5. Carinanos P, Galan C, Alcazar P, Dommguez E. Airborne pollen records response to climatic conditions in arid areas of the Iberian Peninsula. Environ Exp Bot.2004:52:11-22.

6. Krassilov VA, Golovneva LB. Inflorescence with tricolpate pollen grains from the Cenomanian of Tschulymo-Yenisey Basin West Siberian. Rev Paleobot Palynol. 2001;115:99-106.

7. Ollerton J. The evolution of pollinator-plant relationships within the arthropods. Bol SocEntomol Aragon Vol. Monografico. 1999;741-58.

8. Gorelick R. Did insect pollination cause increased seed plant diversity? Biol J Linn Soc. 2001;74:407-27

9. Reyes-Carrillo JL, Munoz-Soto R. Pollen collection in he melon cultivation , vegetation surroundings and curiosities in his harvest by bees (Apis *mellifera* L.) in the Comarca Lagunera . Report of the 10th International Conference on Apfcola Actualization , Tlaxcala, Tlaxcala, Mexico. 2003;24-9.

10. Schmidt RH. The arid zones of Mexico: climatic extremes and conceptualization of the Sonoran Desert. J Arid Environ.1989;16:241-56.

11. Reyes-Carrillo JL, Cabrera-Reyes J, Galarza-Mendoza JL, Orozco-Vidal JA. ^The bees they collect materials alien to pollen during the lean season ? Report of the 14th International Conference on Apfcola Actualization , Boca del Rto , Veracruz. 2007;35-9.

12. Waller GD. A modification of the OAC pollen trap. Am Bee J. 1980;120:119 -21.

13. Steel RGD, Torrie JH. Principles and procedures of statistics. McGraw-Hill Book Company, Inc.New York, Toronto, London. 1960;481p.

14. Desmedt L, Hotier L, Giurfa M, Velarde R, de Brito-Sanchez G. Absence of food alternatives promotes risk-prone feeding of unpalatable substances in honey bees. Sci Rep. 2016;6:31809.

15. Wright GA, Mustard JA, Simcock NK, Ross-Taylor AAR, McNicholas LD, Popescu A, et al. Parallel reinforcement pathways for conditioned food aversions in the honeybee. Curr Biol. 2010;20:2234-40.

16. de Brito-Sanchez MG, Ortigao-Farias JR, Gauthier M, Liu FL, Giurfa M. Taste perception in honeybees: just a taste of honey? Arthropod-PlantInteract. 2007;1:69-76.

17. Nyeko P, Edwards-Jones G, Day R. Honeybee, *Apis mellifera* (Hymenoptera: Apidae), leaf damage on *Alnus* species in Uganda: A blessing or curse in agroforestry? Bull Entomol Res. 2002;92(5):405-12.

18. Reyes-Carrillo JL, Munoz-Soto R, Cano-Rtos P, Eischen FA, Blanco Contreras E. Pollen Atlas of the Comarca Lagunera , Mexico. Guzman Editores , Mexico, DF 2009;347p.

19. Reyes-Carrillo JL, Eischen FA, Cano- Rtos P. Pollen collection for the bees meKferas in cultivated areas of melon and vegetation wild . Report of the 13th International Update Congress Apfcola , San Luis PotosG San Luis PotosG Mexico. 2006; 67-73.

20. Dreller C, Tarpy DR. Perception of the pollen need by foragers in a honey bee colony. Anim Behav. 2000;59:91-6.

21. Davis AR. Further miscellaneous constituents of corbicular pollen loads from *Apis mellifera*: petals, stamens, anther threads, and thrips. ISHS Acta Hortic.1997;437:199-206

22. Maurizio A. Further investigations on pollen panties: Contribution to the recording of pollinating ratios in different areas of Switzerland. [Weitereuntersuchungenanpollenhoschen: beitragzurerfassung der pollentrachtverhaltnisse in verschiedenengegenden der Schweiz. German]. 1953;2(20):485-556.

23. Milne CP, Pries KJ. Honeybees with larger corbiculae carry larger pollen pellets. J Apic Res. 1986;25(1):53-4.

24. Mayack C, Naug D. Starving honeybees lose self-control. Biol Lett. 2015;11(1):20140820.

25. Seeley TD, Camazine S, Sneyd J. Collective decision-making in honey bees: how colonies choose among nectar sources. BehavEcolSociobiol. 1991;28:277-90.

26. Goodwin RM, Perry JH. Use of pollen traps to investigate the foraging behaviour of honey bee colonies in kiwifruit orchards. N. Z. J. Crop Hortic. Sci. 1992;20:23-6.

" bees without food, hives losses "

Saying Spanish

5. Africanization
Rubi Munoz Soto, Jose Luis Galarza Mendoza and Jose Luis Reyes Carrillo
Introduction

A value is estimated at the level world of 212 billion dollars by crop pollination concept facilitated for the bees every year [1]. This digit would result much bigger if you could calculate also he benefit of pollination of wild plants [2].

A problem priority for beekeeping Mexicana , and especially for crop pollination , is the Africanization of bee populations . The bees Africanized are hyphorids of bee races European and African that were created in Brazil in 1956 with the purpose of developing an improvement program genetic . They arrived in Mexico in 1986 when they entered the first swarms across the Guatemala border after 29 years of migration from Brazil . Between the main effects undesirables of bees africanized are his behavior highly defensive and his trend nomad who makes them abandon or escape from the colonies [3,4].

The hyphorids , now bees Africanized , they are less defensive but certainly They attack more frequently than bees European . The beekeepers select hive queens less defensive hives to replace those of the more defensive hives and thus mitigate gradually problem [5]. The bees africanized they can react against an intruder three times faster than bees Europeans and pursue their aggressors to distances greater than a kilometer and then return to their nests staying altered by several days [6]. Apart from its greater willingness to defend itself , the bee Africana is a little smaller , it forms swarms more frequently , its colonies are smaller , it tends to to store less honey and under conditions unfavorable abandon the nest [5.] The evasion or emigration of all the individuals of a colony is a characteristic that bees africanized they express very much frequency . This behavior is due to the fact that these insects are highly susceptible to disturbances caused by beekeepers , predators or natural phenomena , such as noise , handling excessive , heat intense and scarcity of water and food . The evasion of the hives occurs with very little frequency in breed bees European , but in africanized can observe in the hives from 30 to 100 percent [7].

The bees african have his origin in tropical climate zones hot with long periods of drought , and therefore millennia they have faced conditions rustic and difficult that they have made adopt mechanisms to survive ; with his predisposition to emigration in search for resources and their high ability reproductive , the development of behavior defensive sharp , product of being constantly facing a large number of natural enemies in its African habitat including man, and for it to the minor feeling of threat , they respond chasing his aggressor by big distances and in great number [8]. In Mexico there have been presented swarms huge with many queens , who when arriving at a place are divided in little ones nuclei colonizing big surfaces.

Swarm giant africanized landed in a walnut tree in Hermosillo, Sonora (Photography Hector Genaro Galindo Rodriguez)

When he beekeeper performs the extraction of pollen and honey in hives bees Africanized people are disturbed , they feel threatened and can leave the hive [9].
The most used way to determine the differences between bees European and Africanized , both present in beekeeping national , it is known as Method quick to identify bees Africanized (FABIS, *Fast Africanized Bee Identification System*) [10,11]
. This method consider the measurements of the characteristics morphological of the forewings and posterior femurs of a bee sample . The measurement of the anterior wings is called FABIS I, and the measurement of the FABIS II femurs . This is based in he made of bee africanized has a minor size of both wing and femur [12]. Although at the moment exist techniques genetics at the level cell phone very sophisticated techniques to characterize bee populations in commercial hives [13], the FABIS method follows being more favored for this purpose because it is simple , economical and does not require equipment specialized .
Not only the swarms wild they can be Africanized or Africanization process , but also the hives managed by beekeepers present this freak still when he beekeeper Change your queens regularly . Thus result important review periodically he status of the hives and monitor about all he degree of aggressiveness . The defunct National Program for the Prevention and Control of the African Bee reported the first swarm africanized positive in the municipality of San Pedro de las Colonias Coahuila, in October 1991. The purpose of this chapter is to describe the panorama of the presence of the bee africanized in the hives of the Comarca Lagunera .

Studies made since 1992
They were sampled hives and were analyzed swarms captured in the municipalities of Tlahualilo , Bermejillo , Mapimi , Gomez Palacio, Lerdo, Ciudad Juarez, Simon BoKvar and San Pedro del Gallo in he state of Durango, and Torreon, Matamoros, Viesca, Francisco I. Madero and San Pedro de las Colonias in he state of Coahuila.
The place where they took place the analysis for the diagnosis of Africanization of hives technified and swarms was he Beekeeping Laboratory of the Technological Institute Agricultural No. 10, for years 1992 and 1993 [14], in he Biology laboratory of the Autonomous University Agraria Antonio Narro, Laguna Unit for the years 2002, 2003 and 2004-2005 [15] and only in hives technical in 2012 [16] and 2018 [17].
The bees were collected in jars labeled for your ID containing 70% alcohol as conservative and Each one, a minimum of 50 workers were introduced .

Taking samples from hives technical (Photographs Jose Luis Reyes Carrillo)

Identification method FABIS I morphometric

Identification of bees by this method is done measuring the wing length of a batch of 12 bees taken from a random sample and comparing he average obtained with the values critics , the same ones that provide he result and consequent his identification . Through a dissection wearing a watchmaker 's clamp , a total of 12 forewings are detached from the right side of each bee holding firmly to the specimen by he chest . With other clamp comes off the wing from the base, making sure to preserve the notch of the dorsal vein . With the help of the microscope stereoscope the wings are checked , making sure that they are stay in perfect conditions of the edges .
Each lot of 12 wings is placed in rows of six, on a coverslip . To protect them, it is placed other coverslip on top and the edges are joined with tape adhesive , sealing this preparation way . Each one of these preparations are accommodated in a mount plastic for slides , marking with pencil at the bottom the identification data , such as he case number and date of receipt . To see them , they are placed all slides in he carousel of a transparency projector .

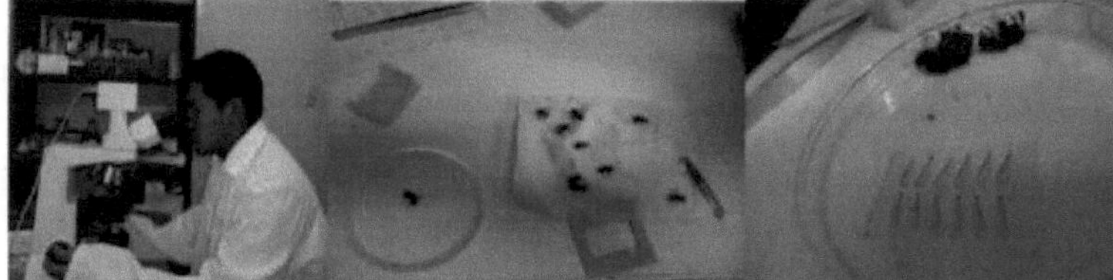
Forewing assembly in laboratory (Photographs Jose Luis Reyes Carrillo)

To make the measurement , a one centimeter ocular micrometer is used . This accessory is for use current in the microscopes and consists in a ruler very small that is used to measure what is observed through the lens . Is taken so he micrometer and mount in a slide frame , just as was done with the wing preparations . It is projected in a screen or smooth wall and adjust the distance of the projector until the image of a 50 centimeter long ruler . Previously, the measurement of 100 centimeters was used , which is equal to 100 times he centimeter that measures he ocular micrometer , but the difficulty of having more screen and projection space , as well how to find and manage one meter rules , led to do better a modification in the formula, allowing So work with half meter rules and multiply he result by two.

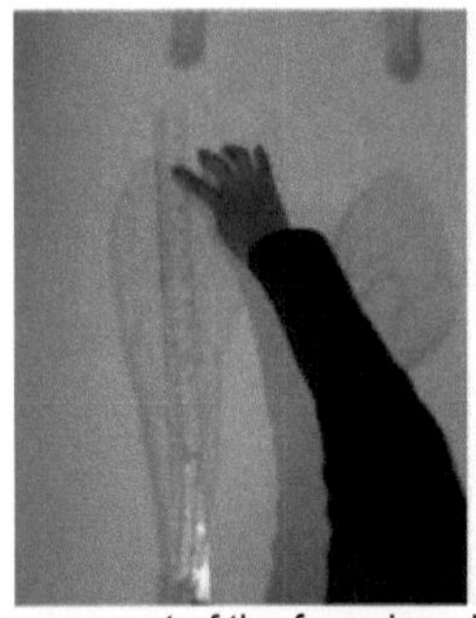 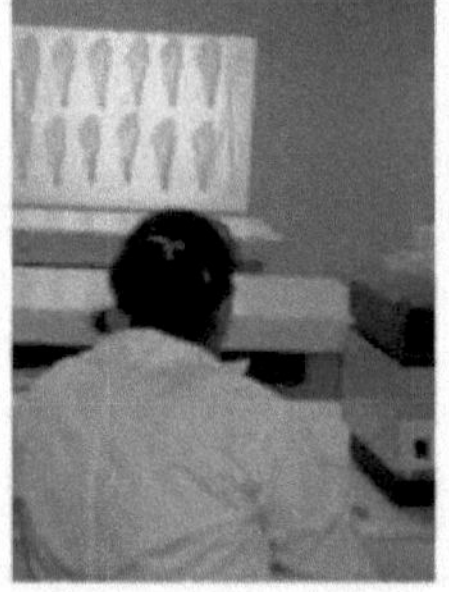 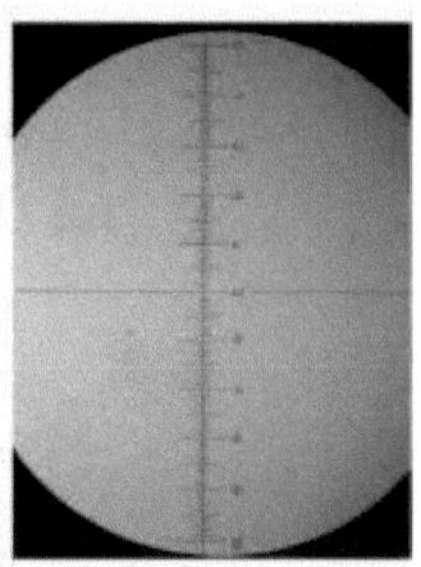

Measurement of the forewings in laboratory and projection of the micrometer to measure it with a 50 centimeter rule (Photographs Rubi Munoz Soto)

Determining the degree of Africanization of bees by he FABIS I method is done measuring the length of 10 of the 12 mounted wings in a slide . This allows discard two wings defects from natural or caused damage during he assembly process . The average of the measurements is compared to the values critics , which is a constant calculated from many studies and measurements of bee wings European and used as Referrer conventional in these cases [12] .

Measurement of the bee femurs according to the FABIS II method in laboratory (Photography Rubi Munoz Soto)

Hives technical

The first year of the study (1992), 5% Africanization was detected in hive bees technical , being the remaining 95% of blood European , and there was no samples suspicious . For him year following (1993) a lower percentage of bees European , an increase in the Africanized ones and 8% of samples appeared suspicious , this is bees that could not be determine with the FABIS method in a category defined . This is common as result of the hybridization process . at the pace of nine years (2002) it was found that bees african were present in half of the colonies sampled , leaving only 42% European and 8% suspicious , this can observe in the next graph :

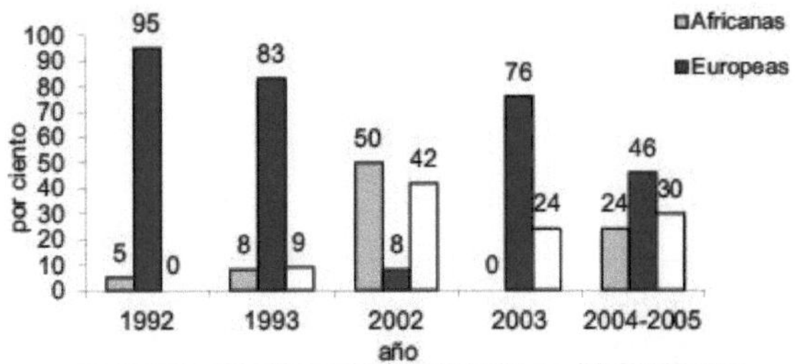

Abejas africanas, europeas y sospechosas en colmenas por el
método FABIS en la Comarca Lagunera 1992-2005.

African , European and suspicious bees in hives by he FABIS method in the Lagunera Region 1992-2005.

Year following (2003) in the hives no bees Africanized . Three quarters (76%) were European and the remaining 24% were classified as suspicious However, for the period 2004-2005, 46% of the hives were European , less than half , 24% Africanized and 30%, almost the third remaining part resulted suspicious
The percentages of bees Africans increased consistently during the next 11 years until reaching 50 %. This is probably because the practice of changing bees queens in the hives that beekeepers they had state practicing regularly and that the presence of Africans decreased decreased in this period . This is reflected in the last year in which only 46 % were European .

Hives technical (photography Jose Luis Reyes Carrillo)

Swarms wild
in swarms wild the situation is different ; In the beginning in 1992 the percentage africanized was high (68%). Since it was detected the first swarm africanized In 1991, Africanization has continued present year with year ; but , yes for practical purposes we consider the samples suspicious with a certain degree of Africanization and we add the percentages , we will find that almost the three quarters of the swarms wild captured were already Africanized . The above is assumed to occur naturally as there is no control over the progeny , nor the swarm to which the bee is most inclined . African

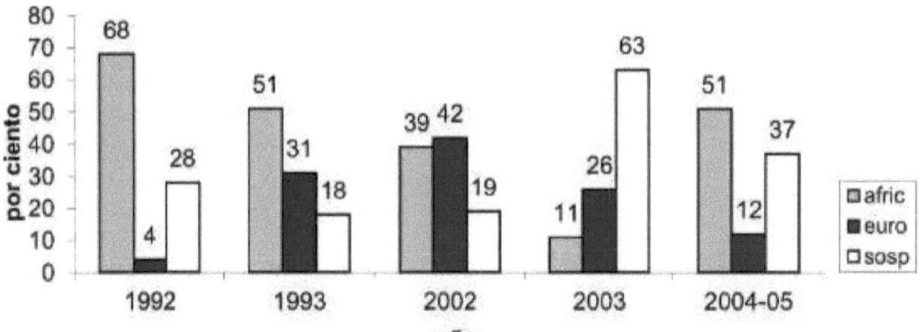

African , European and suspicious bees in swarms by he FABIS method in the Lagunera Region 1992-2005.

Swarm of bees in mesquite (Photography Jose Luis Galarza Mendoza)

When totaling the percentages found bees African , European and suspicious you can see that from the appearance of bees positive for Africanization and suspicious In 1992, his growth immediate was notable in very short time Well, by 1999, 50 % of the bees , swarms and hives already were Africanized .

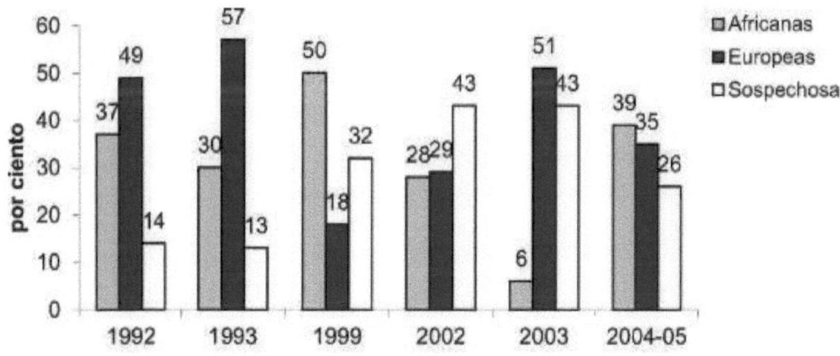

Total bees African , European and suspicious by he FABIS method in the Lagunera Region 1992-2005.

For him year 2002, European and African they had the same amount percentage , but Yeah We consider that the suspicious they can have a degree of Africanization he imbalance is notable in favor of African women .
The year 2003 reflects he effort of the beekeepers to replace their bees queens Well that year presents fewer colonies with Africans but the suspects have a high percentage . For the last period of study , African positive and suspicious Altogether they are two to one in his proportion with bees European , which

61

reflects he Advance invasive in colonies that should be change the queen

Hives technified 2012

For him year 2012, ten years Afterwards , they were only taken bee samples from hives technology in the region, finding that the majority were European , with 20% African and only 9% suspicious :

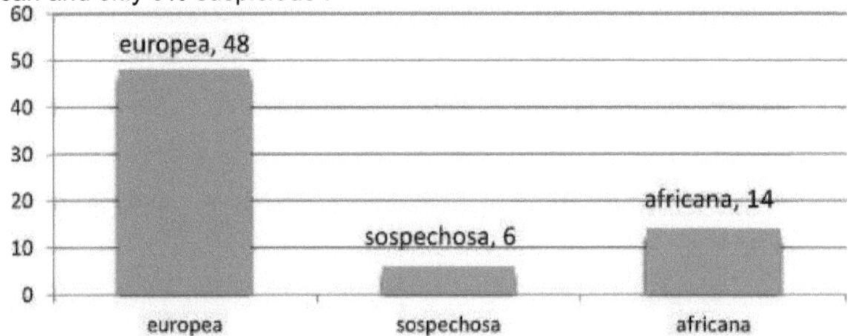

Percentage of overall sampling analyzed by the FABIS I and FABIS II methods in hives of the Lagunera region in 2012

Hives technified 2018

For him year 2018 the hives technical of the beekeepers lagooners showed a percentage dominant (90%) of bees European , 7% bees suspicious and only 3% African . These results lead to the assumption that beekeeping in the region has gone developing with the timely and constant introduction of bees queens European , thanks to the fact that there are three local bee keepers queens , certificates by the Secretary of Agriculture and Rural Development, which provides the beekeepers of the region and the country .

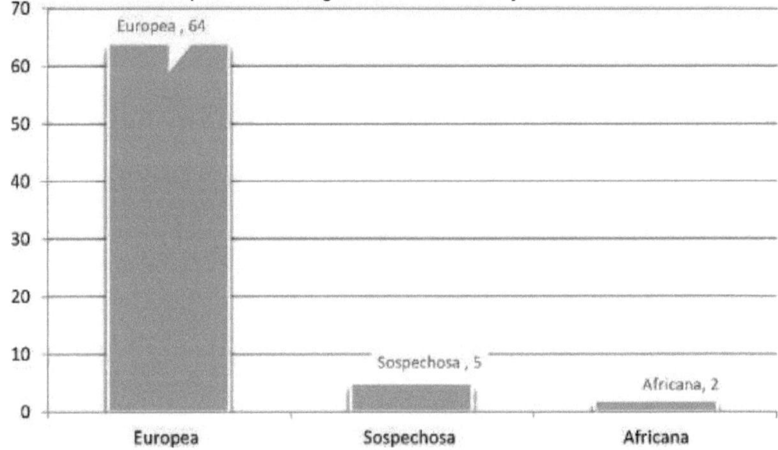

Conclusions

With the presence confirmed bee africanized since 1991 in the Lagunera region can conclude that the hives show so much progress as setbacks in Africanization in function of the effort they make the beekeepers changing queens annually . From his arrival , the swarms wild were africanizing reaching 51 per hundred in he penultimate study period . The number of swarms africanized confirmed as^ as those swarms suspicious , indicate the presence dominant bee Africanized This affects in a way direct in the hives technical either by invasion or by capture of the swarms wild and their encolmenation by part of the beekeeper who take advantage of . The studies revealed that the last years was possible maintain the hives technical with 90% predominance of blood European possibly thanks to the change of the bee queen by the producers .

Capture of a swarm where bees are seen entering the hive (Photography Jose Luis Reyes Carrillo)

References

1.   Contreras-Ramirez DN, Perez-Leon MI, Payro de la Cruz E, Rodriguez-Ortiz G, Castaneda-Hidalgo E, Gomez-Ugalde RM. Behavior defensive , sanitary and production of Apis *mellifera* L. ecotypes in Tabasco, Mexico. Rev Mexicana Cienc Agric. 2016;7(8):1867-77.
2.   Aguero JI, Rollin O, Torretta JP, Aizen MA, Requier F, Garibaldi LA. Impacts of the bee melifera envelope plants and bees wild animals in natural habitats. Ecosystems 2018;27(2):60-9.
3.   Moffett JO, Maki DL, Andrew T, Meliton-Fierro M. The Africanized bee in Chiapas, Mexico. Am Bee J. 1987;127;517–19.
4.   Meliton-Fierro M, Munoz MJ, Lopez A, Sumuano X, Salcedo H, Roblero G. Detection and control of the Africanized bee in coastal Chiapas, Mexico. AmBee J. 1988;128(4):272-75.
5.   Mari-Mut JA. Bees, flowers and honey . Third edition in Spanish . 2018; [Online] https://issuu.com/jamarimutt/docs/abejas__flores_y_miel (Consulted 05/27/20)
6.   Martinez- Puc FJ, Cetzal -Ix W, Gonzalez-Valdivia NA. beekeeping in Campeche: Importance economic and challenges to increase his production . 2018;[Online] http://ru.iiec.unam.mx/3826/W69-Mart % C3%ADnez-Cetzal-Gonz%C3%A1lez.pdf (Consulted 09/16/20)
7.   Guzman-Novoa E, Correa-Benitez A, Espinosa-Montano JG, Guzman-Novoa G. Colonization , impact and control of bees meliferas africanized in Mexico. Vet Mex. 2011;42149-78.
8.   Lovato-Vila I. Origin of the Africanized Bee . The bee murderer : the case that shocked to America. Professional   Magazine   About   Biology .   2017 ;[Online   ] https://allyouneedisbiology.wordpress.com/2017/02/19/abeja-asesina/ (Consulted 05/26/20)
9.   Uribe-Rubio JL, Guzman-Novoa E, Hunt GJ, Correa-Benitez A, Zozaya-Rubio JA. Effect of Africanization on honey production , behavior defensiveness and size of bees meliferas (*Apis mellifera* L.) in the Mexican highlands . Vet Mex. 2003;34(1):47-59.
10.   Rinderer TE, Silvester HA, Brown MA, de Villa J, Pesante D, Collins AM. Field and simplified techniques for identifying africanized and european honey bees. Apidologie 1986;(17):33-48.
11.   Rinderer TE, Allens H, Buco M, Lancaster VA, Herbert EW, Collins AM, Hellmich RL. Improved simple technique for identifying africanized and european honey bees. Apidologie 1987;(18):179-96.
12.   Program for the Prevention and Control of the African Bee (PNPCAA). Methods morphometrics for bee identification . Guidance techniques no. 3 SARH, Mexico. 1991; Printers SA de CV Mexico, DF
13.   Delaney DA, M. Meixner MD, Schiff NM, Sheppard WS. Genetic characterization of commercial honey bee (Hymenoptera: Apidae) populations in the United States by using mitochondrial and microsatellite markers. Ann EntomolSoc Am. 2009;102(4):666-73.
14.   Galarza-Mendoza JL. Bee detection african *Apis mellifera* s cutellata in the Lagunera Region of Coahuila and Durango. Engineer Thesis Agronomist in Production Livestock . technological Institute Agricultural No. 10, Torreon, Coahuila, Mexico. 1996;63p.
15.   Espitia -Villalva S. Bee detection Africanized (*Apis mellifera* scutellata). Veterinary Medical Thesis Zootechnician Autonomous University Agraria Antonio Narro, UL Torreon, Coahuila, Mexico. 2007;60p.
16.   Morales-Perez GE. Bee determination Africana (*Apis mellifera* scutellata) in the Lagunera region . Engineer Thesis in Agroecology . autonomous University Agraria Antonio Narro, UL Torreon,

Coahuila, Mexico. 2012;58p.

17. Hernandez-Salinas CU. Determination of the Africanization of bees (*Apis mellifera* L.) in the Lagunera Region . Engineer Thesis Agronomist in Horticulture . autonomous University Agraria Antonio Narro, UL Torreon, Coahuila, Mexico. 2019;51p.

"The bees that have Honey in their mouth they have bites in the queue"

Proverb Scottish

6. Syndrome of the disappearance or collapse of hives
Azucena Vargas Valero Introduction

The bees best spheres are perhaps those that perform he more predominant role in the activity agriculture , since around a third of the crops They depend directly on them to be pollinated . It is estimated that his contribution to the quality and quantity of crops It is worth approximately 265 billion dollars . by anus . That's where you are bees be key pieces for food safety and maintenance of biodiversity [1] . However, in the last three decades it has been documented a frightening decline in bees meKferas at level world . Numerous studies they have revealed that the causes derive from a multitude of factors that converge in harm to bees [2] ; more than one decrease , the phenomenon has come manifesting as a disappearance of the colonies; one day they are and up to date following They are no longer there . TO are losses , which in addition to affecting the bees also affect seriously the activities apfcultural and agricultural , they have been called collectively as Disappearing Hive Syndrome or Colony Collapse Disorder (CCD) . Is characterized for the absence massive and sudden outbreak of bees from a hive or even thousands of hives . According to the investigations pioneers were moving forward , they left finding more and more causes until adding more than 60 factors associated with the DCC [3] , among the which stand out those related to a feeding poor management practices inadequate , habitat degradation , change climate , the use of pesticides and the presence of various pathogens [4-6] .

The different agents pathogens that have a incidence important about the bees meKfers affect each one in a different way . For example , him mite parasite external varroa affects he system immunological , the behavior , orientation reverberating seriously in the activity of bees foragers , in addition to being a mechanical and biological vector of viruses [7] . Meanwhile , the microsporidium *Nosema* it affects he tract bee digestive adults , causing in the workers a reduction In he gland development hypopharyngeal and minor accumulation of body fat [8,9] . Other parasite is the tracheal mite that lives and reproduces mainly in the bee 's tracheas , affecting his system respiratory , although also can meet in the head and in the sacks air thoracic and abdominal . This organism feeds on the bee 's hemolymph . host and, like varroa, is a vector of several bee viruses [10] .

As for the pesticides , bees are exposed when looking for the nectar- polyferous resources , on all If the colonies are located near agricultural areas [11] . Exposure to doses sublethal pesticides can affect he behavior of the bee 12,13 ' foraging 14,15 ' its longevity 16 ' thermoregulation [17] as well as his learning olfactory and its memory [18] , [19] . pesticide residues they can accrue in he pollen packed in the cells , called bee bread , in honey and in wax [20,21] . The loss of bee colonies by this cause reported in regions of Europe and North America have been greater than 30% [22] , [23] . In our pa^s also exists this serious problem [24] and it he present chapter illustrates the most relevant factors , and based on in cases of the Comarca Lagunera , the main supplier of colonies to pollinate the melon

Importance of bees meliferas as pollinators

Among the great diversity of animals pollinators that exist in he world bees are the primordial ones , and especially the bee meKfera , which is the species predominant and main due to its crop intensive in most of the countries of the world by his easy management , adaptation to different environments, ability to collect nectar and pollen from a wide variety of plants , as well as also , since the economic point of view , the benefits obtained from their products [25-27] . Notwithstanding the above, the most important contribution is the pollination of the vegetation. wild and crops , contributing in this way way to the conservation of biodiversity and ecosystems and to improve food production [28,29] .

The importance of bees for vegetation wild

As we have I come seeing in chapters above , from the perspective of the beings humans , pollination is a service that bees they provide to the ecosystem . Without this function essential through which it is transported effectively he pollen from a flower to another , these would be seen seriously affected [30] Well, 60 to 90% of the vegetation wild depend directly or indirectly from the insects pollinators for reproduction [31] . The natural habitat of the vegetation Wild is essential as it provides bees with nesting sites and foraging areas . about all when the crops agricultural are not in flowering [32] , [33] . The conservation of these areas around agricultural fields increases the diversity and abundance of bees [34] getting better he pollination service .

Bee on the chicalote flower (Photography Juan Cabrera Reyes)

The importance of bees for agriculture

The abundance and diversity of bees they assure a sustained provision of pollination services to a wide diversity of species vegetables and a better food quality of those that are for the human consumption [29] . However, the practices agricultural intensive of the last decades they have devastated big extensions of the natural habitat in addition to poisoning with agrochemicals the fields, putting in danger to some species of bees and causing the inability for them and other pollinators can pollinate properly crops [35] . A viable alternative is systems small agricultural scale , since they can stimulate the pollinators and improve plant reproduction [36] .

A global estimate suggests that the reproduction of more than 87.5% of plants that produce flowers and fruits depends mainly from the insects pollinators like bees [37,38] and a third of the crops agricultural depend on being pollinated by they . For this reason many farmers rent bee hives to complement the activity pollinator of the natural pollinators , achieving a improvement in the production of at least 75 percent [31] . Even though the cereals self -pollinate or are pollinated by he wind , it has been shown that with pollination by insects like bees , these tend to increase the quality and quantity of production [39] ; Likewise , some products horticultural such as nuts , fruits , vegetables , oilseeds and some fodder used to feed livestock would seriously affected by he decline in pollinators [40] .

Cereal cultivation (Photography Jose Luis Reyes Carrillo)

The Disorder of Colony Collapse

Documented losses of bee colonies they have occurred more than 100 years ago . Since 1869 there have been recorded at least 18 episodes of colony loss ; by For example , in 1905 and 1919 90 percent percent of bee colonies died on the Isle of Wight in the United Kingdom . So^ too In 1910 Australia reported 59 per percent losses and by 1975 he reported again high losses . The term used to name the phenomenon was Colony Collapse Disorder (CCD) [3] expression coined in the USA to define a syndrome that is characterized by : 1) the death of part or all of the colony, with the presence of bees dead within or near the hive , 2) the disappearance of part or all of the colony with abandonment of food reserves and offspring and 3) the weakening of the colony characterized due to slow development during the spring even in conditions apparently optimal [41.42] .

The Colony Collapse Disorder was reported in the United States of America by first time in 2006, registering losses higher than the third part of their bee hives [2,3,40,43] . In it continent European showed up a similar situation [44] , as well as also in Canada, China and other parts of Asia [45] , [46] . In Mexico the DCC is considered a problem emerging without statistics officers about the factors that affect or favor he hive collapse . Some studies report the presence of pesticides in samples of honey and wax [47.48] from the hives abandoned For him case of the region of La Comarca Lagunera from 2010 to 2016 was presented a 35% decrease in bee colonies without knowing the causes that caused the decrease , although all the evidence I point out the damage caused by insecticides [24] .

Hives showing he Collapse , the interior with the honeycombs empty , without bees dead and with the presence of food (Photography Jose Luis Reyes Carrillo)

Factors Associated with Colony Collapse Disorder
exist diverse factors biotic or abiotic associated with the DCC, but none of them have been recognized as he only responsible . In fact , DCC is considered to be a problem with multifactorial causes. Between the main factors are found a feeding poor management practices inadequate , destruction or disturbance of the natural habitat, change climate , exposure to pesticides , presence of parasites such as varroasis and nosemiasis , as well as the synergy between all of them [4,6] . Now they present the main :

Destruction o perturbation del habitat
The modification of the landscape through the fragmentation , degradation and destruction of the natural habitat and its replacement by new habitats created by the man, the practices agricultural intensive and large monoculture extensions influence negatively in the plant- pollinator interaction at an individual, population and community scale [31,49] .
Under these circumstances , bees , which are generalists , have greater chances of survival than specialists , dedicated to a genus or species of plant, since the diversity of nectar- pollinare resources provides them adequate nutrition and this they win greater immunity , impacting this made favorably on the health of honeybees [50] .

Climate change
To level world he change climatic is one of the largest threats to biodiversity . The high ones temperatures , droughts , storms , floods , among others events climatic extremes they have caused changes in he plant development and imbalances in plant and pollinator populations . To continue and increase these changes could lead to the extinction of both [29] , [51] .
It is feared he impact that the increase of the events climatic extremes could have in the local pollinator communities of each region of the globe , such as in bees [49] , since when they move the flowering patterns , coming early or late , desynchronized he cycle annual bee and these lose he flowering period of vegetation of importance as food source . It is estimated that in the near future between 17 and 50 % of species pollinators they will suffer food shortage due to disturbances in the flowering patterns of plants [52] .

Typical semi- desert vegetation lagoon (Photography Juan Cabrera Reyes)

Presence of pesticides

Without place to any doubt , it is known that pesticides they have been the main responsible in the mortality of pollinators , mainly bees [53] . Are remain exposed when they come out in search for nectar , pollen and guttation water - which consists in the secretion of droplets of substances Liquid and sugary from the interior of the plant to the surface of the leaves or stems -, on all If the colonies are located near agricultural areas that have been sprayed with pesticides [11] .

The impacts of the agrochemicals about the bees vary according to conditions environmental . The casualties temperatures and a diet low in proteins They increase the susceptibility of bees to poisoning [54,55] . The effects sublethal doses pesticide drops in bees they are diverse , for example some studies they have proven that they can affect he behavior of the bee [12] , [13] , foraging 1 , [15] , its longevity [16] , thermoregulation [17] as well as his learning olfactory and its memory [18,19] .

pesticide residues they can accrue in the bee products , such as bee bread , which is pollen packed in the cells , honey and wax [20,21] . The latter has a high residual pesticide storage capacity [56] . in a honeycomb contaminated , waste they can be transferred to honey , posing a risk not only to bees but for the consumers . It is crucial here consider that honey is consumed in diverse presentations and for various purposes, both domestic as commercial and even industrial : " honey in honeycomb ", honey as food additive , in he fruit treatment , such as supplement food or as a flavoring [57] .

The most used pesticides in agriculture and therefore those with the greatest presence in the interior of the hive are the herbicides , acaricides and fungicides . Those who have been marked as the main causes of bee mortality are insecticides of the neonicotinoid family 58,59 , which were marketed by first time in 1991. Its use grew up very quickly and became in the most used insecticides in he world ; again his popularity began to increase from the year 2000 , registering in 120 countries with more than 140 agricultural uses [60] . These insecticides Synthetics are derived from nicotine , to which they owe his name . They have a mode of action similar to that of nicotine , with the disadvantage that while nicotine degrades easily , the neonicotinoids have a high persistence in he atmosphere . The classification of these insecticides gone varying between generations , being used different terms . They are currently classified in three different generations [60,61] .

dying bees at the entrance and falling to the ground in front of the hive (Photography Octavio Vazquez Calvete)

insecticides neonicotinoids are neurotoxins that time in the inside of the insect body interfere in his system central nervous system blocking the transmission of impulses nervous This leads to a quick modification of your behavior , paralysis and later death . Everything happens in nails few hours [60,62] .

Your acceptance in the market is due to its broad spectrum in insect control suckers as the aphids , flies white , spiders , some microlepidoptera , coleoptera and termites , among others . Their action is systemic , they are powerful at low dosage , provide long- term control and have a pronounced residual activity . Another advantage that they offer and that has been responsible for your success , it's your versatility in the application forms , since this can be do through irrigation any way: by drip , in hydroponics or spraying foliar . It also can apply straight to the ground in granules , for injection in trunks , by dispersion of oils and how seed treatment forming a coating layer about [60] grain .

At the moment are prohibited by the Commission European although temporarily due to the growing worry by he risk they pose to bees meliferas and in general for everyone pollinators [63] . In Mexico they are free to use without restriction some and they exist presentations for use domestic for the control of cockroaches and garden pests . An estimated consumption of 55,000 tons of pesticides annually in Mexico, of the which house and garden represent 8 by hundred .

Other characteristic is that they are effective systemic , that is , when applied they do not remain in the outside of the plant but are absorbed into the vascular tissue . This is attractive from the point of view of the producer since it offers protection against insects herbivores , but it is highly harmful to bees , especially when arrive at the plant in searches for nectar or pollen , since the pesticide this present in all parts and products of the plant [64] . Another major drawback is that, depending on the application method and type of crop , only between 1.6 % and 28 % of the ingredient active is absorbed by cultivation [65] while most of it disperses in the environment , the soil , the water and the air . Based on This has been found to be neonicotinoids represent a risk significant for water surface waters and the diversity of aquatic and terrestrial fauna [62,66] .

Barely recently at the level world it has finally started to be provided attention to the effects that these insecticides have about the insects pollinators . This has raised controversies due to the contradiction between studies that have performed to prove that bees cannot be seen affected by this type of compounds [67,68] while others show that their high toxicity has effects lethal or sublethal . It is necessary highlight that there is other effect associated with the neonicotinoids , the which consists in he synergism that arises between pesticides and diseases , such as nosemiasis , which means a sum of agents harmful substances that weaken the colonies and make them more vulnerable , reaching to cause he collapse of hives [69] . The effects synergistic of the pesticides also they have been registered with others parasites such as varroa and some viruses [2] . All this , combined between two or more of these factors those that affect significantly he system immune system of bees , influencing in your reserves energetic and doing them highly susceptible to disorientation and death [70] is what has caused he collapse of hives [71] .

Presence of pathogens

The relationship between the presence of pathogens and the Colony collapse is not well defined , but research they have revealed a high pathogen prevalence in collapsed colonies that have lost a part or all of the bee population [72] . Between the main pathogens are found he varroa mite and the Nosema intestinal parasite .

Varroa destructor

Varroosis or varroasis is an external parasitosis caused by he mite *Varroa destructor* that affects brood and bees adults . This parasitosis causes large losses economical in beekeeping and is considered as he problem apfcola most important healthcare system worldwide [70,73,74] .

Cycle organic varroa

Varroa is characterized by present two phases in his life cycle , phase phoretic and reproductive . The phase foretica is the displacement stage during which adult females mount in the bees adults parasitizing and using them as a means of dispersion. The phase reproductive happens when the female mites ovoposit inside the cells where the babies develop . The males and the different nymphal stages are short duration and only develop inside the capped cells [75] .

Varroa spreads and transmits by contact direct between bees adults . Could so tell yourself that your cycle biological starts when a bee adult parasitized by the mites that feed on your hemolymph enters in contact with a bee from another hive and the mite moves . It is not at all strange that in sometimes the bees foragers lose the route and end finding and entering in other hive , this is known as drift . The strait contact between bees inside the hive favors he mite Oh infesting many bees , including bees Nurses , which by feeding the larvae , facilitate the entry of female mites into the cells shortly before they be sealed for metamorphosis - 15 to 20 hours before in worker cells and 40 to 50 hours before in drone cells - . Once it penetrates in a cell containing a bee larva , the mite dives to the background , where it is located he larval food , thus protecting itself from the action of cleaning and removal of agents strangers by part of the bees workers cleaners . The infestation rate is 8 to 10 times higher in drone farming . in comparison with the breeding of workers [76] . How are you plague is relatively new for the bee meKfera it has not developed a behavior defensive efficient yet .

When the infestation this already course a female mite usually deposit between four and five eggs on the larvae of workers with a birth ratio of one male and three to four females. When it's on a drone breeding , which are preferred by the Mites by have a gestation period longer than that of the workers , the female mite will lay 5 to 7 eggs , with a birth ratio of one male and five to six females. The cycle complete from egg to adult lasts approximately six to seven days , reaching even in sometimes maturity reproductive still inside the cell , before it is opened for the alleged bee hatching [75] . The feasibility reproductive performance of varroa progeny depends on the Mites decendents reach he stadium adult and manage to mate [76] .

Females can be fertilized even since before leaving the cell and being willing immediately to infest others cells or to stick to bees adults through which they can spread and infest other colonies. A female mite can have one and a half cycles reproductive in average in conditions normal , and can get to live from two to eight months in the interior of the hive . The parasite population increases quickly and exponentially when increases the amount of breeding in the hive because it favors its reproduction [73].

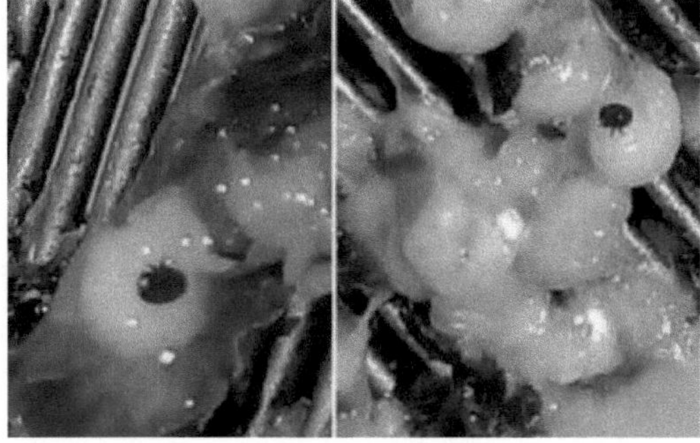

drone larvae parasitized by varroa (Photography Hector Genaro Galindo Rodriguez)

Transmission mechanisms

In addition to drift - the entry of bees into hives that are not their own -, a form of dispersion or transmission of varroa between hives that already exist. we mentioned , the contagion also happens by factors As the pillage , the unrestricted entry of drones to different hives or swarm invasion . The practices inappropriate handling , such as he exchange of combs with brood and bees between infested and healthy colonies , the introduction of queens accompanied by workers infested and the hive movement very infested to apiaries with low infestation levels are factors additional contagion that depends entirely from the management of the beekeeper [73].

Damage caused for the varroa

Varroa parasitosis causes a notable decline in the hives , due to the weakness they suffer many bees adults by the Mites parasites that suck their hemolymph , reducing their the body protein levels . The bees parasitized exhibit behavior abnormally nervous , loss of ability in he flight , difficulty in learning capacity , advancement of foraging age , absences prolonged periods of the colony and some They don't come back anymore [77]. Besides this type of parasitosis causes a state of immunosuppression in the bee that makes it vulnerable to the action of others agents biological , mainly viruses, such as the one that causes Deformed Wings , the paralysis virus acute Israel , paralysis virus acute and pouching virus 75 ·

When there are levels very high levels of infestation , Mites consume also he tissue bee fat adults , increasing his vulnerability to certain pesticides [78,79]. This may be a factor in colony collapse [80] when Bush to a large number of bees and brood , causing a decline sudden population adult , malformations in the workers remaining , breeding skipped , alteration of capacity reproductive , delay in he replacement generational , increase and association with other diseases [79]. Without the application of a treatment effective and timely against infestation, most colonies collapse in a period of 2 to 3 years [40].

Nosemiasis

The intestinal parasite *Nosema apis* and especially *Nosema ceranae* are known coparceners in he hive collapse . It is a pathogen widely extended to level worldwide , causing numerous losses economical . It wreaks its havoc both individually and in association with others agents that affect mainly bees foraging adults [9],[81] . in the bee Eastern meKfera has greater pathogenicity [81] .

Cycle biology of nosemiasis

The infection by *Nosema* start through ingestion of spores through food contaminated like honey and pollen , which when reaching the stomach of the bees germinate by effect of digestive juice [8] . The spores are microscopic , oval and measure approximately between 4 and 6 microns long by 2 to 4 microns wide [82] .

When the spores they mature , they are released in he intestine to carry out the reproduction and germination of new spores . At that moment they can infect to others cells epithelial and proliferate in the bee or go outside with the feces . To finish he cycle biological , the spore has to be under conditions optimal temperatures , during the 48 to 60 hours after the start of the infection . Temperature is an important factor for the germination and multiplication of spores , which can be see limited by below 20 degrees centigrade and by above 35 degrees centigrade . The optimal temperature for your development and multiplication is between 30 and 35 degrees Celsius [8] .

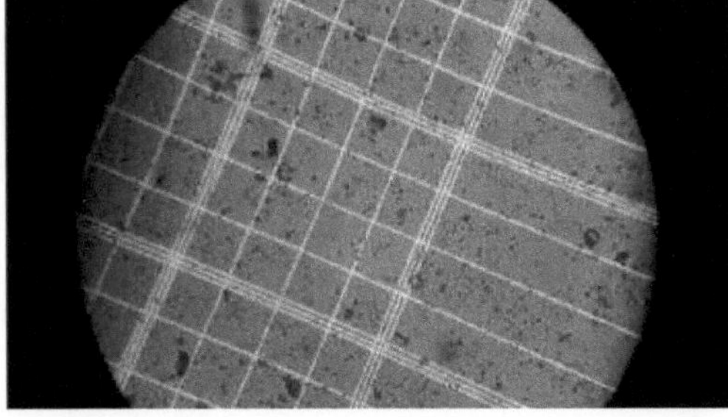

Microscope image of thousands *of* Nosema spores showing the slide grid for your counting (Azucena Vargas Valero Photography)

Transmission mechanisms

The spores of the pathogen are transmitted indirectly when a healthy bee enters in contact with material or food contaminated with them . Some of the ducts or infection factors are management practices inappropriate as he exchange unnoticed movement of drawers , floors , or racks from a sick colony to a healthy one since the waste Fecal waste from bees is a means of transmission , the introduction of queens or workers infected and pillage of an infested colony to a healthy colony [8,83] . The form of transmission direct is less frequent and occurs through trophallaxis , which is the food exchange between bees ; both workers as queens and drones are sensitive equally [8] .

Damage caused per *Nosema*

The symptoms observed in the bees adults are paralysis , trembling wings , inability to fly, loss of thorax pubescence , spots fecal inside and outside the hive , dysentery and occasional groups of bees sick in he ground or in the pit , which generally present a swollen abdomen . In the workers some of the effects additional are reduction in he gland development hypopharyngeal , fat reduction corporal , inability to take advantage of pollen , consumption accelerated protein stored in his body and minor longevity . In queens the presence of nosema spores causes besides a decrease in egg laying [8] . A queen can stay infected during a lot time , so it is recommended carry out he change of queens every year [84] .

Hive showing the stains typical excrement by bee diarrhea due to nosemiasis (photograph Juan Ocon Cisneros)

Acariasis tracheal

acariasis tracheal or acarapisosis is a parasitosis caused by he mite microscopic *Acarapis woodi* , who stays in the tracheas of bees adults , in the head and in the sacks air thoracic and abdominal . This guest undesirable , it feeds on the hemolymph , carries viruses and causes important economic losses [10] .

Cycle biology of acariasis tracheal

The bees young people are infested by he female mite when they enter in contact physical with bees parasitized older than 14 days of age . It travels between the outer pubescence of the bee 's thorax sick and from there it goes to the bee young man , holding on with the help of his nails . Subsequently , and guided by air currents produced by the bee 's breath host , find he hole of a trachea through which penetrates The fertilized female mite enters in the bee 's trachea young and at four or five days puts five to seven eggs that take a while in open between three and six days . The adults mate in the inside of the tracheae and fertilized females they can give place to the next generation in the same trachea or they leave it to infest to other bees [76] .

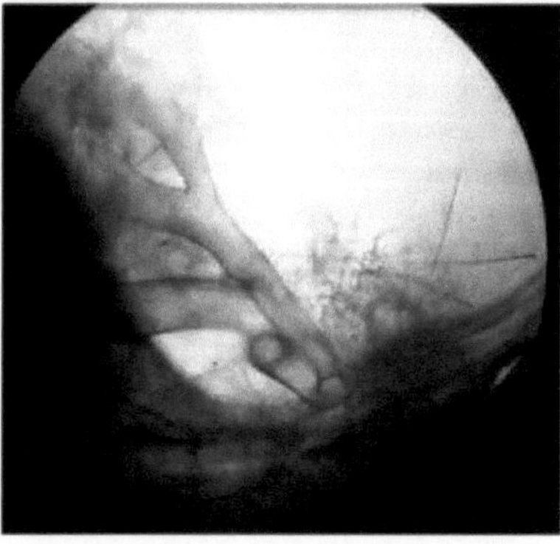

Tracheas healthy bees observed under the microscope . They can be distinguish the spiracles respiratory , tube - shaped , clean and translucent free from the mite (Photography Luda Marcial Salvador)

Both the nymphs as the adults of these Mites feed on the bee 's hemolymph . They suck it from the walls of the trachea , piercing it with the help of their fangs . This some lesions that later become pigmented and that, because it is a characteristic typically used for diagnosis in he laboratory.The accumulation of mites in the bee 's trachea provokes a insufficient supply of oxygen to the flight muscles . This explains why bees lose ability to fly, in addition to the general weakening that results from the effect of toxins produced by the parasites and loss of hemolymph . The life span of a bee disease is approximately 30% shorter than that of a healthy bee [73] .

Transmission mechanisms

The most frequent way the disease arrives in an apiary healthy in areas free of the problem it is the migration of swarms or the introduction of bees queens and workers sick.The Mites are not able to survive without a live host for more than 12 hours and they infestation levels rise after long periods of confinement when in contact nearby the bees [73] . parasitize bees young up to six days old . Older bees are immune to mite penetration . in their tracheas , possibly by he hardening of the villi surrounding the openings of the first pair of tracheae thoracic by where penetrates the parasite [76] . The transmission is seen favored by factors as poor management by part of the beekeeper , the bee entry pillagers or in swarm drift , capture and hivement infected .

Damage caused for acariasis tracheal

The signs clinical signs of acariasis cannot always be observe and are generally only evident when the Colony infestation levels are greater than 50 % . Among the clearest manifestations the wings are " dislocated " in many bees that flap their wings without being able to fly, distended abdomen , bees dead or dying in front of the piqueras . If you look at a bee infected carefully , it will be noted that the chest this already devoid of hairiness , so it looks black and shiny. For another side is very notorious that bees sick lose he sting instinct . This last feature is not only characteristic neither exclusive to this problem , since it can also observe in cases of starvation , poisoning by insecticides , poisoning by food consumption fermented , changes abrupt in the temperature environmental , or cases to present others diseases as Nosemosis , amoebiasis and paralysis by viruses [73] .

Conclusions

Colony Collapse Disorder (CCD), which causes the loss massive bees meKferas , and in consequence of thousands of hives , has Causes multifactorial as feeding poor management practices inadequate , habitat reduction , change climate , the use indiscriminate use of pesticides and the presence of pathogens . Pesticides they kill the bees and doses sublethal they can affect seriously his foraging behavior , its longevity, its thermoregulation , as well as his learning olfactory and memory . The pathogens affect bees in different ways , altering he system immunological , the behavior , orientation , activity of bees foragers , their longevity and survival , increasing the damages when they combine all these factors .

Bee carrying pollen clusters in the corbiculae of its legs rear (Photography Jose Omar Enriquez Santacruz)

References

1.   Lautenbach S, Seppelt R, Liebscher J, Dormann CF. Spatial and temporal trends of global pollination benefit. PLoS One. 2012;7(4)e35954.
2.   Brutscher LM, McMenamin AJ, Flenniken ML. The Buzz about honey bee viruses. PLoS Pathog . 2016;12(8):1-7.
3.   VanEngelsdorp D, Evans JD, Saegerman C, Mullin C, Haubruge E, Nguyen BK, et al. Colony collapse disorder: A descriptive study. PLoS One. 2009;4(8)e6481.
4.   Staveley JP, Law SA, Fairbrother A, Menzie C. A causal analysis of observed declines in managed honey
bees (*Apis mellifera*). Hum Ecol Risk Assess. 2014;20(2):566-91.
5.   DeGrandi-Hoffman G, Chen Y. Nutrition, immunity and viral infections in honey bees. Curr Opin Insect Sci. 2015;10:170-6.
6.   McMenamin AJ, Genersch E. Honey bee colony losses and associated viruses. Curr Opin Insect Sci. 2015;8:121-9.
7.   Yang X, Foster D. Effects of parasitization by *Varroa destructor* on survivorship and physiological traits of *Apis mellifera* in correlation with viral incidence and microbial challenge. Parasitology. 2007;134(3):405- 12.
8.   Fries I. *Nosema apis* -a parasite in the honey bee colony. Bee World 2015;74(1):5-19.
9.   Higes M, Martm -Hernandez R, Botfas C, Bailon EG, Gonzalez-Porto AV, Barrios L, et al. How natural infection by *Nosema ceranae* causes honeybee colony collapse. Environ Microbiol . 2008;1-11.
10.   Garrido-Bailon E, Bartolome C, Prieto L, Botfas C, Martinez-Salvador A, Meana A, et al. The prevalence of *Acarapis woodi* in Spanish honey bee (*Apis mellifera*) colonies. Exp Parasitol . 2012;132(2012):530-6.
11.   O'Neal ST, Anderson TD, Wu-Smart JY. Interactions between pesticides and pathogen susceptibility in honey bees. Curr Opin Insect Sci. 2018;26:57-62.
12.   Balbuena MS, Tison L, Hahn ML, Greggers U, Menzel R, Farina WM. Effects of sublethal doses of glyphosate on honeybee navigation. PLoS One. 2015;10(10):2799-805.
13.   El Hassani AK, Dacher M, Gauthier M, Armengaud C. Effects of sublethal doses of fipronil on the behavior of the honeybee (*Apis mellifera*). Pharmacol Biochem Behav. 2005;82(1):30-9.
14.   Cresswell JE, Thompson HM. Comment on "A common pesticide survival in honey bees". Science. 2012;337:1453.
15.   Henry M, Beguin M, Requier F, Rollin O, Odoux J, Aupinel P, et al. A common pesticide decreases foraging success and survival in honey bees. Science. 2012;336:3-5.
16.   Wu JY, Anelli CM, Sheppard WS. Sub-lethal effects of pesticide residues in brood comb on worker honey bee (*Apis mellifera*) development and longevity. PLoS One; 2011;6(2):e14720.
17.   Tosi S, Demares FJ, Nicolson SW, Medrzycki P, Pirk CWW, Human H. Effects of a neonicotinoid pesticide on thermoregulation of African honey bees (*Apis mellifera* scutellata). J Insect Physiol. 2016;93:56-63.
18.   Williamson SM, Wright GA. Exposure to multiple cholinergic pesticides impairs olfactory learning and memory in honeybees. J Exp Biol. 2013;216(10):1799-807.
19.   Lu C, Warchol KM, Callahan RA. Sub-lethal exposure to neonicotinoids impaired honey bees winterization before proceeding to colony collapse disorder. Bull Insectology. 2014;67(1):125-30.
20.   Johnson RM, Ellis MD, Mullin CA, Frazier M. Pesticides and honey bee toxicity - USA. Apidologie. 2010;41(3):312-31.

21. Lozano A, Hernando MD, Ucles S, Hakme E, Fernandez-Alba AR. Identification and measurement of veterinary drug residues in beehive products. Food Chem. 2019;274:61-70.

22. vanEngelsdorp D, Meixner D. A historical review of managed honey bee populations in Europe and the United States and the factors that may affect them. J Invertebr Pathol. 2010;103:580-95.

23. Chauzat MP, Jacques A, Laurent M, Bougeard S, Hendrikx P, Ribiere-ChabertM, et al. Risk indicators affecting honeybee colony survival in Europe: one year of surveillance. Apidologie. 2016;47(3):348-378.

24. Vargas-Valero A, Reyes-Carrillo JL, Gaspar- Rairurez O , Moreno-Resendez A. Parasitosis and pesticide residues in honey and wax in bee colonies . Ecosist Recur Agropec . 2021;8(2):e 2827. DOI: 10.19136/ era.a 8n2.2827

25. Saturni FT, Jaffe R, Metzger JP. Landscape structure influences bee community and coffee pollination at different spatial scales. Agric Ecosyst Environ. 2016;235:1 -12.

26. Martins KT, Gonzalez A, Lechowicz MJ. Pollination services are mediated by bee functional diversity and landscape context. Agric Ecosyst Environ. 2015;200:12-20.

27. Gaines-Day HR, Gratton C. Crop yield is correlated with honey bee hive density but not in high-woodland landscapes. Agric Ecosyst Environ. 2016;218:53-7.

28. Jones PL, Agrawal AA. Learning in insect pollinators and herbivores. Annu Rev Entomol. 2017;(62):53- 71.

29. Food and Agriculture Organization of the United Nations (FAO). The importance of bees and other pollinators for food and agriculture. FAO. 2018. [En lmea] http://www.fao.org/3/I9527EN/i9527en.pdf (Consulta 01/08/20).

30. Garantonakis N, Varikou K, Birouraki A, Edwards M, Kalliakaki V, Andrinopoulos F. Comparing the pollination services of honey bees and wild bees in a watermelon field. Sci Hortic. 2016;204:138-44.

31. Kremen C, Williams NM, Aizen MA, Gemmill-Herren B, LeBuhn G, Minckley R, et al. Pollination and other ecosystem services produced by mobile organisms: A conceptual framework for the effects of land-use change. Ecol Lett. 2007;10(4):299-314.

32. Minarro D, Garcia R, Martmez -Sastre R. Insects pollinators in agriculture : importance and management of its biodiversity . Ecosystems . 2018;27(2):81-90.

33. Simba LD, Foord SH, Thebault E, van Veen FJF, Joseph GS, Seymour CL. Indirect interactions between crops and natural vegetation through flower visitors: the importance of temporal as well as spatial
spillover. Agric Ecosyst Environ. 2018;253:148 -56.

34. Alomar D, Gonzalez-Estevez MA, Traveset A, Lazaro A. The intertwined effects of natural vegetation, local flower community, and pollinator diversity on the production of almond trees. Agric Ecosyst Environ. 2018;264:34-43.

35. Klein AM, Vaissiere BE, Cane JH, Steffan-Dewenter I, Cunningham SA, Kremen C, et al. Importance of pollinators in changing landscapes for world crops. Proc Biol Sci. 2007;274:303-13.

36. Hass AL, Kormann UG, Tscharntke T, Clough Y, Fahrig L, Martin J, et al. Landscape configurational heterogeneity by small-scale agriculture, not crop diversity, maintains pollinators and plant reproduction in western Europe. Proc R Soc B. 2018;(285):20172242.

37. Ollerton J, Winfree R, Tarrant S. How many flowering plants are pollinated by animals? Oikos. 2011;(321):321-6.

38. Ollerton J. Pollinator diversity: distribution, ecological function, and conservation. Annu Rev Ecol Evol Sist. 2017;48:353-76.

39. Nicole W. Pollinator power: nutrition security benefits of an ecosystem service. Environ Health Perspect. 2015;123(8):210-15.

40. Spivak M, Mader E, Vaughan M, Euliss NH. The plight of the bees. Environ Sci Technol. 2011;45(1):34-8.

41. Underwood RM, vanEngelsdorp D. Colony Collapse Disorder: Have we seen this before? Bee Cult.
2007;35(717):13-8.

42. Simon-Delso N, Martin GS, Bruneau E, Minsart LA, Mouret C, Hautier L. Honeybee colony disorder in crop areas: The role of pesticides and viruses. PLoS One. 2014;9(7):1-16.

43. Brutscher LM, Daughenbaugh KF, Flenniken ML. Antiviral defense mechanisms in honey bees. Curr Opin Insect Sci. 2015;10:71-82.

44. Faucon JP, Mathieu L, Ribiere M, Martel AC, Drajnudel P, Zeggane S, et al. Honey bee winter mortality in France in 1999 and 2000. Bee World. 2002;83(1):14-23.

45. Van der Zee R, Pisa L, Andonov S, Brodschneider R, Charriere JD, Chlebo R, et al. Managed honey bee colony losses in Canada, China, Europe, Israel and Turkey, for the winters of 2008-9 and 2009-10. J Apic Res. 2012;51:100-14.

46. Van der Zee R, Brusbardis V, Charriere J, Chlebo R,Coffey MF, Dahle B, et al. Results of international standardised beekeeper surveys of colony losses for winter 2012-2013: analysis of winter loss rates and mixed effects modelling of risk factors for winter loss. J Apic Res. 2014;53(1):19-34.

47. Perez-Pacheco A. Identification of toxic residues in honey from different sources in the central zone of the State of Veracruz. Rev Iberoam Cienc Biol Agropecu. 2012;1(2):1-42.

48. Valdovinos-Flores C, Alcantar-Rosales VM, Gaspar-Ramirez O, Saldana-Loza L, Dorantes-Ugalde

JA. Agricultural pesticide residues in honey and wax combs from Southeastern, Central and Northeastern Mexico. J Apic Res. 2017;56(5):667-79.

49. Goulson D, Nicholls E, Botias C, Rotheray EL. Bee declines driven by combined stress from parasites, pesticides, and lack of flowers. Science. 2015;347(6229):1-11.

50. Alaux C, Ducloz F, Crauser D, Le Conte Y. Diet effects on honeybee immunocompetence. Biol Lett. 2010;6(4):562-65.

51. Bellard C, Cleo Bertelsmeier, Paul Leadley WT,Courchamp F. Impacts of climate change on the future of biodiversity. Ecol Lett. 2012;15:365-77.

52. Memmott J, Craze PG, Waser NM, Price MV. Global warming and the disruption of plant-pollinator interactions. Ecol Lett. 2007;10(8):710-7.

53. Hladik ML, Vandever M, Smalling KL. Exposure of native bees foraging in an agricultural landscape to current-use pesticides. Sci Total Environ. 2016;542:469-77.

54. Archer CR, Pirk CWW, Wright GA, Nicolson SW. Nutrition affects survival in African honeybees exposed to interacting stressors. Funct Ecol. 2014;28(4):913-23.

55. Schmehl DR, Teal PEA, Frazier JL, Grozinger CM. Genomic analysis of the interaction between pesticide exposure and nutrition in honey bees (*Apis mellifera*). J Insect Physiol. 2014;71:177-90.

56. Benuszak J, Laurent M, Chauzat MP. The exposure of honey bees (*Apis mellifera*; Hymenoptera: Apidae) to pesticides: Room for improvement in research. Sci Total Environ. 2017;587-588:423-38.

57. Wilmart O, Legreve A, Scippo ML, Reybroeck W, Urbain B, de Graaf DC, et al. Residues in beeswax: A health risk for the consumer of honey and beeswax? J Agric Food Chem. 2016;64(44):8425-34.

58. Botfas C, Sanchez-Bayo F. Papel de los plaguicidas en la perdida de polinizadores. Ecosistemas. 2018;27(2):34-41.

59. Ostiguy N, Drummond FA, Aronstein K, Eitzer B, Ellis JD, Spivak M, et al. Honey bee exposure to pesticides: A four-year nationwide study. Insects. 2019;10(1):1-34.

60. Jeschke P, Nauen R, Schindler M, Elbert A. Overview of the status and global strategy for neonicotinoids. J Agric Food Chem. 2011;59(7):2897-908.

61. Nauen R, Jeschke P. The neonicotinoid insecticides. *In*: Insect control. Biological and synthetic agents. Jamestown Road, London, UK. Academic Press. 2010;61-114.

62. Anderson JC, Dubetz C, Palace VP. Neonicotinoids in the Canadian aquatic environment: A literature review on current use products with a focus on fate, exposure, and biological effects. Sci Total Environ. 2015;505: 409-22.

63. Brandt A, Gorenflo A, Siede R, Meixner M, Buchler R. The neonicotinoids thiacloprid, imidacloprid, and clothianidin affect the immunocompetence of honey bees (*Apis mellifera* L.). J Insect Physiol. 2016;86:40- 7.

64. Douglas MR, Tooker JF. Large-scale deployment of seed treatments has driven rapid increase in use of neonicotinoid insecticides and preemptive pest management in U.S. field crops. Environ Sci Technol. 2015;49(8):5088-97.

65. Robin SUR, Stork A. Uptake, translocation and metabolism of imidacloprid in plants. Bull Insectology. 2003;56(1):35-40.

66. Morrissey CA, Mineau P, Devries JH, Sanchez-Bayo F, Liess M, Cavallaro MC, et al. Neonicotinoid contamination of global surface waters and associated risk to aquatic invertebrates: A review. Environ Int. 2015;74:291-303.

67. Cutler GC, Scott-Dupree CD. Exposure to clothianidin seed-treated canola has no long-term impact on honey bees. J Econ Entomol. 2007;100(3):765-72.

68. Chauzat AM, Carpentier P, Martel A, Cougoule N, Porta P, Lachaize J, et al. Influence of pesticide residues on honey bee (Hymenoptera: Apidae) colony health in France. Environ Entomol. 2009;38(3):514-23.

69. Van der Sluijs JP, Simon-Delso N, Goulson D, Maxim L, Bonmatin JM, Belzunces LP. Neonicotinoids, bee disorders and the sustainability of pollinator services. Curr Opin Environ Sustain. 2013;5(3-4):293-305.

70. Dainat B, Evans JD, Chen YP, Gauthier L, Neumann P. Predictive markers of honey bee colony collapse. PLoS One. 2012;7(2). e32151.

71. Sanchez-Bayo F, Goka K. Pesticide residues and bees - A risk assessment. PLoS One. 2014;9(4).e94482.

72. Smith KM, Loh EH, Rostal MK, Zambrana- Torrelio CM, Mendiola L, Daszak P. Pathogens, pests, and economics: Drivers of honey bee colony declines and losses. Ecohealth . 2013;10(4):434-45.

73. Guzman-Novoa E, Correa- Benrtez A. Pathology , diagnosis and control of the principal diseases and pests of bees meKferas . Image Publishing House , Mexico, DF. 2012;165p.

74. Meixner MD, Francis RM, Gajda A, Kryger P, Andonov S, Uzunov A, et al. Occurrence of parasites and pathogens in honey bee colonies used in a European genotype-environment interactions experiment. J Apic Res. 2014;53(2):215-19.

75. Rosenkranz P, Aumeier P, Ziegelmann B. Biology and control of *Varroa destructor* . J Invertebr Pathol . 2010; 103: S96-S119.

76. Secretariat for Agriculture, Livestock , Rural Development, Fisheries and Food (SAGARPA). Manual of

pathology apfcola . National bee control program African General Coordination of Livestock . Mexico. 2010;59p.

77. Fuchs S, Kralj J, Brockmann A, Fuchs S. The parasitic mite *Varroa destructor* affects non-associative learning in honey bee foragers, *Apis mellifera* L. J Comp Physiol. 2007;193:363-70.

78. Le Conte Y, Mondet F. Natural selection of honeybees against *Varroa destructor*. *In*: Vreeland RH, Sammataro D (eds.) Beekeeping - From science to practice. Springer International Publishing AG. 2017;189-94.

79. Fries I, Ansen HH, Mdorf AI, Osenkranz PR. Swarming in honey bees (*Apis mellifera*) and *Varroa destructor* population development in Sweden. Apidologie. 2003;34:389-97.

80. Di Prisco G, Annoscia D, Margiotta M, Ferrara R, Varricchio P, Zanni V, et al. A mutualistic symbiosis between a parasitic mite and a pathogenic virus undermines honey bee immunity and health. Proc Natl Acad Sci. 2016;113(12):1-6.

81. Paxton RJ, Klee J, Korpela S, Fries I. *Nosema ceranae* has infected *Apis mellifera* in Europe since at least 1998 and may be more virulent than *Nosema apis*. Apidologie. 2007;38(6):558-65.

82. Ptaszynska AA, Borsuk G, Mutenko W, Demetraki J. Differentiation of *Nosema apis* and *Nosema ceranae* spores under Scanning Electron Microscopy (SEM). J Apic Res. 2014;53(5):537-44.

83. Rangel J, Baum K, Rubink WL, Coulson RN, Johnston JS, Traver BE. Prevalence of *Nosema* species in a feral honey bee population: a 20-year survey. Apidologie. 2016;47(4):561-71.

84. Simeunovic P, Stevanovic J, Cirkovic D, Radojicic S, Stanisic L, Stanimirovic Z. *Nosema ceranae* and queen age influence the reproduction and productivity of the honey bee colony. J Apic Res. 2014;53(5):545-54.

" What is not good for the hive cannot be good for the bees "

Marcus Aurelius

7. Pollination of the melon crop

Veronica Garda-Mendoza and Pedro Cano -Rios

Introduction

The melon is a fruit that contains in its interior a long history and some mysteries , beginning for its origins , since it is not known with certainty his place exact origin , at least there is no consensus among experts yet . In what s^ the majority of the authors is in his origin African , and in consider India as he domestication center , since it is where exists the greatest variability genetics of the species . China, Afghanistan and Spain are considered as centers secondary diversification , due to the important diversity genetics that are in those regions. Something that is perfectly documented is the great development that I experienced his cultivation and the great acceptance that always had among Egyptians , Greeks and Romans , who knew and equally They praised the virtues of this refreshing fruit . His arrival to Spain was somewhat later thanks to the Arabs and from there it spread to the rest of Europe and finally to America [1,2] .

Melon generalities

The melon is a herbaceous stem plant creeping exist diverse varieties , such as Spanish , yellow, written or reticular, Piel de Sapo and some that are known as cantaloupe varieties . The production of cantaloupe melon is one of the most common in national and international markets [3] .

Scar in the separation zone of the melon fruit when it separates from the stem (Photography Jose Luis Reyes Carrillo)

Classification taxonomics
Scientific name *Cucumis* melo L.
Kingdom: Plantae
Subkingdom : Tracheobionta
Division: Magnoliophyta
Class: Magnoliopsida
Subclass : Dilleniidae
Order: Cucurbitales
Family: Cucurbitaceae

Subfamily : Cucurbitoideae
Tribe : Benincaseae
Subtribe : Benincasinae
adaptation geographical
The melon is an annual , herbaceous plant . creeping or climbing . Grown in different areas of the world fundamentally in climates warm and semi-warm . Depending on latitude in which it is grown the melon will show a diversity greater or lesser morphology of both its seeds as well as its fruits , that is , their colors , shapes , thicknesses , resistance of the shell and the same vary . fruit , duration of the cycle and the profile agronomic . The cycle phenological from sowing to fruiting varies from 9o to 110 days [6,7] .

botanical description
The melon is a species that belongs to the Cucurbitaceae family . In this botanical family are found crops such as pumpkin, sand^a , gherkin or cucumber, zucchini , chayote, bitter gourd , cucumber angolo or zocato , estropajo and guiro (bule, gourd or dried gourd) [8] . Among the main cucurbitaceae , the melon has more botanical varieties and also has the fruits with more diverse shapes . This is one consequence of diversity genetics in this species . The fruits of some varieties have better aroma, pulp color , flavor and they are juicier . Most of the melons have plants up to 15 meters long; however, they have developed some cultivars modern with internodes shortened , appearance bushy and performance concentrated . All the Melons are sensitive to frost , but in contrast , many differ in his ability to survive to high environments temperature .

Sexual expression is controlled by factors genetics , as well as also by environment ; at least four factors , such as energy luminica , the photoperiod , the water supply and temperature , have a great influence in this . Normally , the conditions physiological that favor he increased carbohydrates inside the plant, like the ground floor temperature , little available nitrogen , photoperiod short and tall accessibility to moisture , promote female sexual expression . These factors environmental affect he hormonal balance of the plant [4] .

Cycle vegetative
The cycle vegetative this ruled mostly by the temperature . Cano and Gonzalez [9] found that 1,178 units are needed heat -critical point below 10 °C and above 32 °C - to give start to harvest and a total of 1,421 units heat to finish he cycle . This is illustrated in he following chart .

Units heat for each stage phenological through which it passes he melon cultivation

Phenological Stage	Heat Units
Sowing	0
Emergency	48
1st Sheet	120
3rd Sheet	221
5th Sheet	291
Gwa Home	300
Home of Male Flower	382
Home Hermaphrodite Flower	484
Beginning of Fruiting	534
Walnut Size	661
% Fruit Size	801
% Fruit Size	962
% Fruit Size	1142
Harvest Start	1178

Ra^z

The root system of the melon is varied due to diversity genetics among different genotypes . In conditions adequate soil , fertilization and free of any type of stress , most observed type of root is the called triangular, with a main root from which they come roots sides that are decreasing in size towards the tip . However, also appear morphologies typically rectangular , with roots laterals that are distributed with similar length along the main root , being more frequent this guy in the subspecies agrestis , as much as in those of type wild or exotic cultivated , from China, Japan and India [10] .

Structure of the melon root . Left: triangular structure , predominant in the cultivars of the melo subspecies . Right : rectangular structure typical of the entries of the subspecies agrestis [10] .

Stem

The melon plant is creeping or climbing , if it is provided he medium appropriate . Your guides They reach 3 to 4 meters in length , with stems rounded smooth or grooved , provided with a soft hair and simple tendrils . The main stem branches in its base giving place three or four branches or stems secondary . Subsequently , both the main stem and the secondary , they develop new smaller branches or stems [8] .

An argument

The leaves are wider at the base than at the top and are egg-shaped, almost as kidney , usually angular -of five angles -, in occasions with three to seven shallow pointed lobes rounded . They have petioles 4 to 10 centimeters long. The leaf, 8 to 15 centimeters in diameter , is hairy with somewhat wavy -toothed to almost entire edges [8] .

Leaves and stems creeping branches with melon fruits (Photography Veronica Garda Mendoza)

Flower

The flowers are yellow and there are three. types : male , female and hermaphrodite -flowers that present the same time the organs masculine and feminine -. Depending on the presence of these flowers on the plant, they are classified in :

Monoecious : the plant bears male and female flowers Andromonoecious : the plant bears male and hermaphrodite flowers Gynomonoecious : the plant has female and hermaphrodite flowers Trinomonoecious : the plant has the three types of flowers.

Generally the plants are andromonoecious . The male flowers appear before hermaphrodites in groups of three to five flowers in the guide knots primary and never where is it located a flower hermaphrodite Female or hermaphrodite flowers appear lonely in the guide knots high schools . Female flowers are distinguished from male flowers. in he bulge in its base, which is where it is located he ovary . Melon plants produce more male flowers than hermaphrodites . Fertilization is mainly by insects [11] .

Most plants bear male and hermaphrodite flowers , with a greater proportion of male flowers than hermaphrodites . The genotypes of the melons guy yellow , honey dew and japanese have floors bearers of male flowers and hermaphrodites only , while in the Cantaloupe, Charentais , Galia and Harper melons there are plants with male and hermaphrodite flowers and with male and female flowers . The distribution of the flowers depends on the genus of the plant and the flowers themselves . The male flowers appear before hermaphrodites in groups of three to five flowers sprouting in the guide knots primary and never where is it located a flower female or hermaphrodite [11,12] .

hermaphrodite flower in the part left and one male in the part right (Photography Olga Araceli Zapata Ramos)

Fruit

The fruit of the melon has pulp fleshy and leathery shell . The fruits of the melon are diverse in as for your size , color, shape, and texture of the bark , depending on the variety of the plant. Its size varies from the four centimeters long -melon variety agrestis - up to 200 cm -melon variety flexuosus -, and can register a weight that ranges between 50 and 1,500 grams - a size variation of 30 times -. The color of your pulp It can be orange , pink, green , white , or even a mix of these colors , and the color of its shell can be white , cream, cream- greenish , yellow pale to dark , yellow -brown, yellow greenish , green , orange , red, gray or a mix of these colors . The texture of its shell has also many variants : smooth opaque or uniform shiny , with a cover of a rough or corky , warty , striped , reticulated , rough or any other layer combination of them . In it bark case reticulated , the true external color is observed in the spaces exposed among the reticulum threads .

In the melons reticulates that are cultivated in the Lagunera Region , the fruit detachment is naturally when this this already mature. That is because in this variety , like in some others, the Fruits come off the plant when they mature , due to the formation of a separation zone . at the union of the base of the peduncle ; in others melons this does not happen [8] , as is the case of the melons Harper type . In those cases the fruits remain joined firmly to stem even after they are mature for not possessing this characteristic [4] . It is common that in some varieties are noticeable along the fruit from 9 to 12 ribs separated by sutures forming a net more or less dense Regarding the form, the fruits They can be spherical or ellipsoid . The fruits soften as they ripen . Some varieties develop in its inner essences aromatic scented , although others remain almost odorless

Composition of the fruit

Most of the content of the fruit is water and in his maturity reaches a solids content soluble between 7 and 12 °Brix. The melon contains a large amount of water , which in some varieties reaches up to 92 % and a 6% lower amount of sugar than others fruit ; fact that together with the fact that hardly contains fat , makes melon one of the fruits with the lowest content caloric Your content in carbohydrates is easy assimilation , contributes a amount appreciable in various vitamins and minerals . The hundred grams of pulp , provide almost half the dose daily recommended vitamin C , and along with oranges , it is one of the fruits with the highest content in vitamin B. It is noteworthy also his content in provitamin A, mainly beta carotene , which is transformed in vitamin A in our organism . The wealth in these carotenes increases in the melons with more orange flesh . As for the minerals , highlights his wealth in potassium . It is a food restorative that promotes activity physical and intellectual , since the potassium improvement he functioning of muscles and nerves ; and together with sodium regulates the water balance in he organism , and normalizes he rhythm cardiac . It also contains quantities appreciable phosphorus , iron and magnesium , so melon is a natural remineralizing product .

It is one of the richest fruits in mineral sodium. The high degree of water in this fruit stimulates the kidneys so that they function more efficiently , facilitating the elimination of waste substances and toxins , and improving kidney function . Also this indicated in dehydration states accompanied by mineral losses by diarrhea , sweating abundant and febrile seizures [13] .

Seeds

In it center of the melon fruit is located a large number of seeds . The fruit part is consumed, while the seeds are used as waste materials . It has been found that the seeds are constituted by moisture , fat , protein , fiber , ash and carbohydrates

, depending on the variety is the percentage of which are formed. It has also been found that the seeds are formed due to high oil content he which has potential to become in a new source of edible oil since it contains a high level of acids polyunsaturated fatty acids [14].

Composition of different melon seeds

Component	Composition approximate (%)		
	melon var. Toilet [14]	Imbrido melon AF-522 [15] !	melon var. Saccharinus [16]
Humidity	4.9	7.78	6.0
Grease / Oil	25.0	30.83	32.3
Protein	25.0	14.91	19.3
Fiber	23.3	19.00	
Ash	2.4	4.20	3.9
Carbohydrate	19.8	22.94	-

Nutritional value of the fruit

The melon is delicious in vitamins important , like riboflavin , thiamine and acid folic It is also a good source of pro- vitamin A and vitamin C. Researchers they have found that the concentrations of solids soluble , sucrose , sugars total , в — carotene , and the 5-methyltetrahydrofolic acid varies in different parts of the fruit [14].

The most important carbohydrate in the melons cross-linked is sucrose . This accumulates in the last 10 to 12 days before harvest . The fruit does not contain starch or other carbohydrate reserve ; by Therefore , if it is harvested early , the fruit will not be appropriately sweet. In it following table shows the composition nutritional value of 100 grams of cantaloupe melon , although the quantities will vary depending on the variety and its quality .

Nutrient composition of 100 grams of the edible part of melon [17]

Input per serving		Minerals		Vitamins	
Energy (Kcal)	55.44	Calcium (mg)	15.6	Vit. B1 Thiamine (mg)	0.05
Protein (g)	0.88	Iron (mg)	0.35	Vit. B2 Riboflavin (mg)	0.01
Hydrates carbon (g)	12.40	Iodine (mg)	0.55	Eq. Niacin (mg)	0.66
Fiber (g)	0.73	Magnesium (mg)	11.8	Vit. B6 Pyridoxine (mg)	0.06
fat (g)	0.10	Zinc (mg)	0.29	Ac. Folico (pg)	2.70
AGS (g)	0.03	Selenium (pg)	0.5	Vit. B12 Cyanocobalamin	0.00
AGM (g)	0.01	Sodium (mg)	17.0	Vit. C Ac. ascorbic (mg)	32.10
AGP (g)	0.02	Potase (mg)	310.0	Retinol (pg)	0.00
AGP/AGS	0.58	Phosphorus (mg)	0.00	Carotenoids (and carotenes pg)	669.3
(AGP/AGM)/AGS	1.08			Vit. A Eq. Retinol (pg)	111.9
Cholesterol (mg)	0.00			Vit. D (pg)	0.00

Water (g)   85.90

Split melon in half showing the cavity that contains the seeds (Photography Veronica Garcia Mendoza)

Importance international

Over ten years the world production of melon has been in average of 26,180,803 million tons in a surface harvested from 1,073,371 hectares with a yield average of 24.4 tons/ha:

World melon production (2009-2018) [18]

Year	Production (ton)	Harvested Area (ton/ha)	Yield (ha)
2018	27,349,214	26.1	1,047,283
2017	26,624,465	25.3	1,051,105
2016	26,563,103	24.6	1,080,066
2015	25,537,777	24.6	1,037,414
2014	26,059,468	24.6	1,058,209
2013	26,238,525	24.7	1,060,792
2012	25,797,647	23.9	1,077,222
2011	25,839,056	23.5	1,100,084
2010	25,967,660	23.2	1,119,754
2009	25,831,116	23.4	1,101,782

The highest production of melon is in he continent Asian with 73% of the production worldwide , followed by a participation of 13.6% by he American continent ; he continent European contributes 7 % , continent African with 5.5% and Oceania with 0.8% [18] :

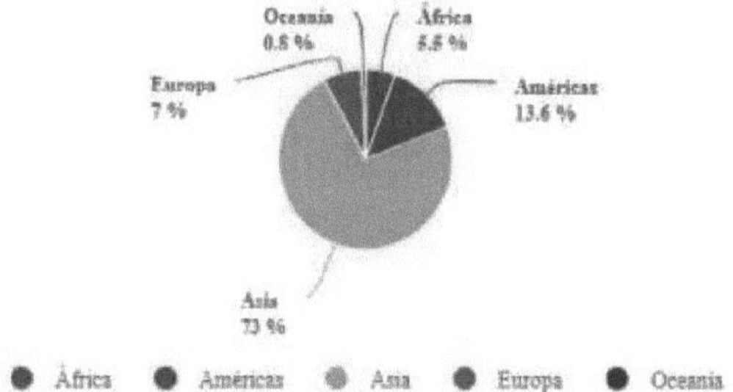

Participation of melon production in he world in he year 2018 [18] .

More than half of the production melon world cup was generated in China in 2018, whose farmers obtained from their exploitations a total of 12.7 million tons , 46.5 percent percent of the world total . The second place in the classification world of melon producers occupies Turqrna , with 1.7 million tons and 6.41 percent cent of the total. Iran, with 1.7 million tons (6.33%) is the third place of the greater melon producers . The fourth position this busy for India, which produces 1.2 million tons , 4.5 percent cent of the total. It follows in fifth place Kasakhastan with 0.89 million tons (3.27 %), United States with 0.87 million tons (3.19%), Egypt with 0.70 million tons (2.56 %) , Spain with 0.66 million tons (2.43 %) , Guatemala with 0.62 million tons (2.28%), Italy in tenth place , with 0.60 million tons (2.22%) , leaving Mexico in eleventh place with 0.59 million tons with a 2.17 percent percent of production melon world cup :

Classification of the main countries producers of mel6 n (2018) 18

	Production	Yield ton/ha	Harvested Area (ha)

	(ton)			
1 China	12,727,263	35.9		354,496
2 Turkey	1,753,942	22.3	78,694	
3 Iran	1,731,443	20.4	85,000	
4 India	1,231,000	22.8	54,000	
5 Honesty	893,857	21.3	42,004	
6 EU	872,080	28.5	30,590	
7 Egypt	701,071	27.1	25,842	
8 Spain	664,353	34.9	19,025	
9 Guatemala	623,405	21.4	29,176	
10 Italy	607,970	24.9	24,400	
11 Mexico	594,608	31.4	18,959	
12 Brazil	581,478	24.9	23,324	
13 Morocco	500,823	30.0	16,716	
14 Afghanistan	329,241	9.4	35,055	
15 Honduras	293,234	49.8	5,891	
16 France	255,100	19.0	13,407	
17 Bangladesh	227,000	19.3	11,736	
18 Australia	224,672	28.9	7,786	
19 Venezuela	205,179	19.8	10,370	
20 South Korea	167,638	41.4	4,051	
21 Japan	143,078	22.7	6,316	
22 Costa Rica	143,015	32.2	4,437	
23 Indonesia	118,708	17.5	6,773	
24 Iraq	113,538	10.7	10,634	
25 Greece	103,585	21.6	4,785	
26 Ukraine	102,750	5.8	17,700	
27 Tunisia	102,546	10.5	9,729	
28 Azerbaijan	94,668	15.4	6,130	
29 Argentina	81,861	14.5	5,628	
30 Colombia	75,448	15.5	4,852	
Rest	1,084,660	13.6	79,777	
world	**27,349,214**	**26.1**		**1,047,283**

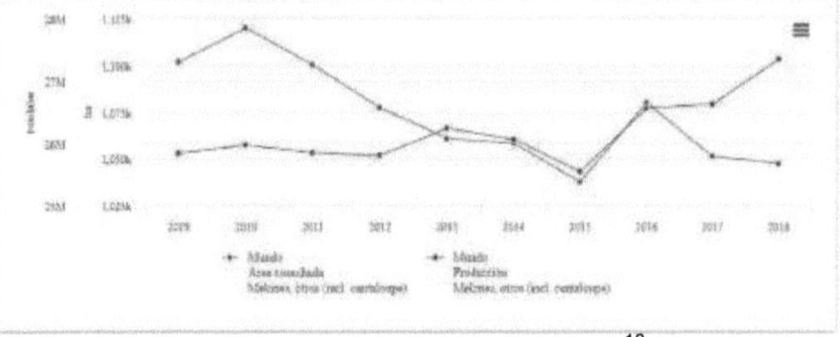

Production melon world and harvested area (2009-2018) [18]

In it year 2018 at the level 27,349,214 million tons of melon were produced worldwide , over a surface of 1,047,283 million hectares , according to the last FAO data :

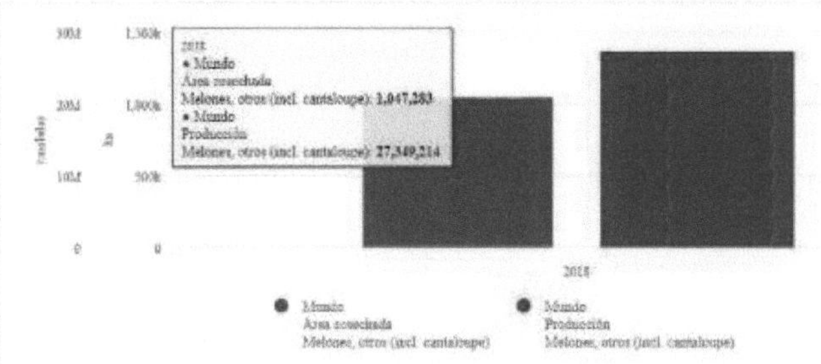

Melon production at level world and harvested area in he year 2018 [18] .

exist different tastes and preferences of consumers according to traditions cultural of each country ; by For example , most of the exports to the United States by part of Costa Rica, corresponded to harper/cantaloupe melon (both types together , since no data is available by separated) with 85%, followed by Honey Dew melon (15%). In contrast , in that year exports to Europe were led by Yellow melon (58%), followed by Harper/Cantaloupe melon (39%) and Galia melon (3%) [19] . The offer in the melon export market requires assess recent cultivars creation , including those of interest to new markets. It is necessary investigate the more appropriate factors of production , and develop strategies commercials to address potential markets to countries like Korea, Japan or Singapore , which are of great importance , either as fresh or frozen product [20] :

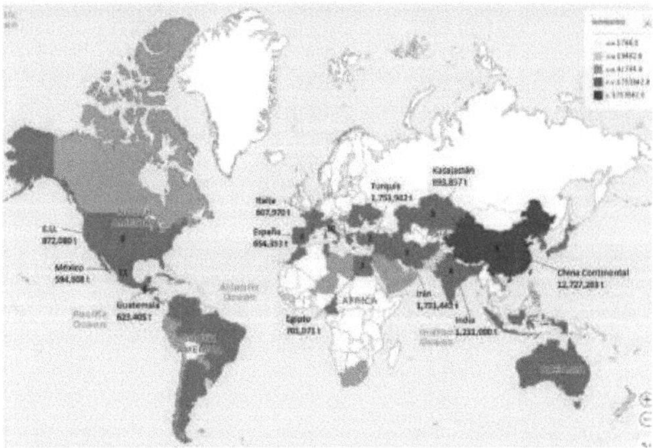

Classification of the main countries melon producers in 2018 [18] .

Importance national

The melon, from the years twenty , it has been a product currency generator for countries , sources of employment and profit income for the producers Mexicans and it is from the years sixty when his presence take importance among producers , derived from a greater demand from both the national market as well as international . During the last Seventy - five years , the Mexican melon has maintained his stake in the international market by his quality . In addition to the spill economics that represents in cultivation areas , the result of labor required for your handling , packaging and marketing , is the third product agricultural in he line of foreign exchange acquisition [21] .

Melon agriculture has developed widely , existing high level technology for your production , which has increased the returns , as degree that the varieties creoles They have disappeared from the market. exist many varieties available , that adapt and give results in the different regions where it is grown , in he country the most important is the Cantaloupe, or Chinese melon and in minor proportion Honey Dew or smooth melon . Of the latter practically everything is exported . With the advances technological have definite the optimal production and quality in the different regions , with November to April produce the southern and pacific states and from May to October those of the Comarca Lagunera , existing in he marked quality fruit at a low price during all he anus . The most important states By the area of melon planted they are: Coahuila, Guerrero, Michoacan, Sonora and Durango [22] .

The production national in he In 2019, it reached 627,135.31 tons : five most important states In the production of melon, the following contributed figures in he 2019 year : Coahuila contributed 25 % of production national melon, which represents a production of 154 thousand tons , Sonora produced 20 % of the national total , that is contributed 124 thousand tons of melon, Guerrero produced a total of 102 thousand tons , that is , it produced 16 % of the national total , Michoacan obtained a production of 86 thousand tons of melon, which represents 14 % of the national total and Durango contributed 9 % of the national total , which is equivalent to 58 thousand tons of melon according to data published by he Information service Agri-Food and Fisheries (SIAP), of the Secretariat of Agriculture and Rural Development (SADER), and the Information System Agri-Food Consultation (SIACON) [22] .

Production national melon (2009-2019)

Ano	Production (tons)	Performance (tons/ha)	Harvested Area (ha)
2019	627,135.31	31.61	19,837.74
2018	594,608.35	31.36	18,958.83
2017	605,134.17	30.92	19,572.80
2016	593,717.15	29.62	20,047.06
2015	561,891.31	28.91	19,435.98
2014	526,990.47	28.79	18,306.69
2013	561,952.87	28.73	19,561.44
2012	574,212.85	28.53	20,126.42
2011	556,026.80	26.97	20,617.65
2010	559,116.03	26.53	21,075.97
2009	547,326.65	26.65	20,541.10

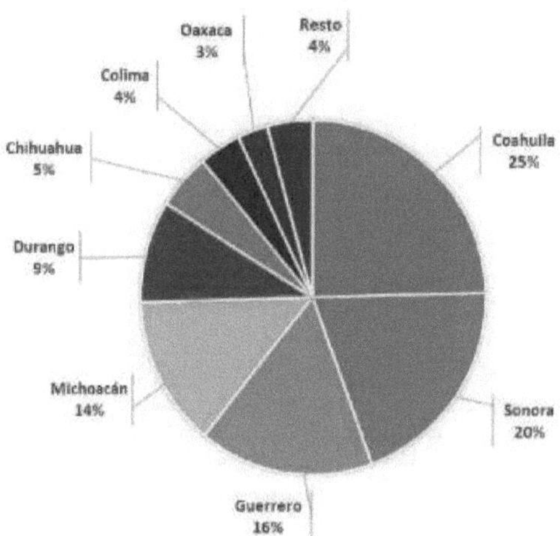

Participation of the state in the production national melon in he year 2019 [22] .

National melon production (2019) [22]			
	Production (ton)	Harvested Area (ton/ha)	Yield ([ha])
1 Coahuila	154,004.99	34.44	4,472.16
2 Sonora	124,062.70	37.26	3,330.00
3 Guerrero	102,955.93	29.20	3,526.45
4 Michoacán	86,344.30	33.78	2,556.00
5 Durango	58,159.70	31.55	1,843.20
6 Chihuahua	32,089.32	28.62	1,121.19
7 Colima	25,050.10	45.75	547.5
8 Oaxaca	17,881.62	15.39	1,161.75
9 Jalisco	6,097.31	24.25	251.41
10 Guanajuato	4,948.04	22.54	219.5
Resto	15,541.30	19.22	808.58
Total	627,135.31	31.61	19,837.74

States with the highest melon production in Mexico [22] .

Regional importance

One of the main reasons why melon cultivation is shown particular interest is because it is a generator of jobs and income for the producers and consequence of foreign exchange for countries that produce it [23].

The Comarca Lagunera , region located in he north of the country , which this integrated by five municipalities of Coahuila: Torreon, Matamoros, San Pedro, Francisco I. Madero and Viesca and ten of Durango: Gomez Palacio, Lerdo, Tlahualilo , Mapimn , Nazas , Rodeo, San Pedro del Gallo, San Luis del Cordero, San Juan de Guadalupe and Simon BoHvar and is characterized for being the main melon region in the country in some months of the year , since the planted areas it has represent about 20% of the national surface [24].

The main melon producing municipalities in the Laguna de Durango region are Mapimi and Tlahualilo , concentrating 25 % of the total melon production obtained in the Laguna Region in 2019, which was 52,108 tons , while in the Laguna de Coahuila region it was obtained a production of 131,286 tons, which represents 60 % of the production of the municipalities with the largest area harvested record , and stand out for being the main melon producers , Viesca, Matamoros and Parrasy can observe it in the following tables and figures :

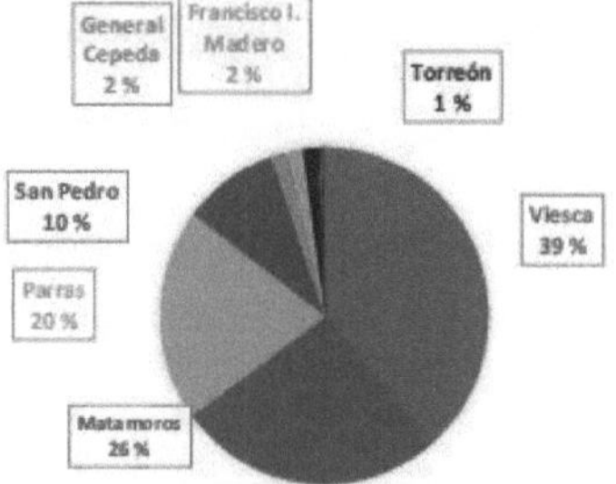

Participation of municipalities in production melon state in the Laguna - Coahuila region
in he year 2019 [22].

Production melon annual in the Laguna Region - Coahuila (2019) [22]

	Production (ton)	Yield (ton/ha)	Harvested Area (ha)
viesca	59,460	35.6	1,672.4
Matamoros	40,627	41.6	976.8
Grapes	31,198	30.3	1,030.0
St. Peter's	14,580	28.3	515.5
General Cepeda	2,416	30.2	80.0
Francis I. Wood	2,340	45.0	52.0
The tower	2,085	36.0	58.0
Zaragoza	455	11.4	40.0
Gentleman	352	14.1	25.0
Border	210	17.5	12.0
Four Cienegas	191	27.3	7.0
Saint Bonaventure	54	27.3	2.0
Lamadrid	33	22.3	1.5

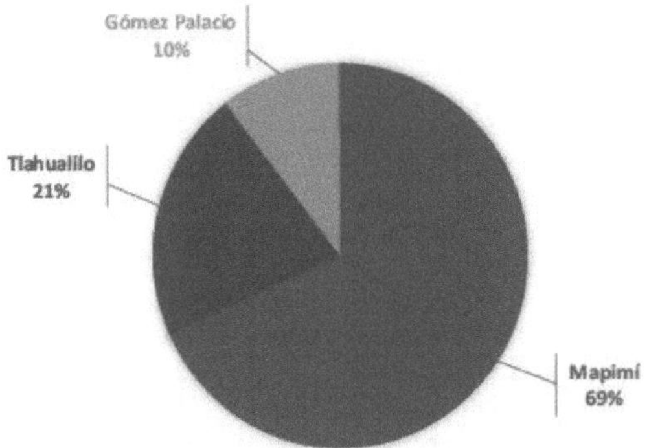

Participation of municipalities in production melon state in the Laguna - Durango region in he year 2019 [22] .

Production melon annual in the Laguna Region - Durango (2019) [22]

	Production (ton)	Yield (ton/ha)	Harvested Area (ha)
Mapimn	39,759	31.3	1,270
Tlahualilo	12,349	28.9	427
Gomez Palacio	6,021	41.5	145
Dull	30	30.0	1

An aspect that producers are giving each increasingly important , and of which they are informing and certifying , it is in he safety aspect . Consumer concern does necessary put in ring bells to encourage good practices agricultural and hygiene in both production as in he packaging and transportation to obtain products harmless high quality .

Varieties

From time back An attempt has been made to standardize the classification of the different varieties of melon, one of the most accepted classifications is the proposal by Naudin in 1859; the following are some modifications or revisions [25] . In the next figure are shown some main types of melons varieties .

92

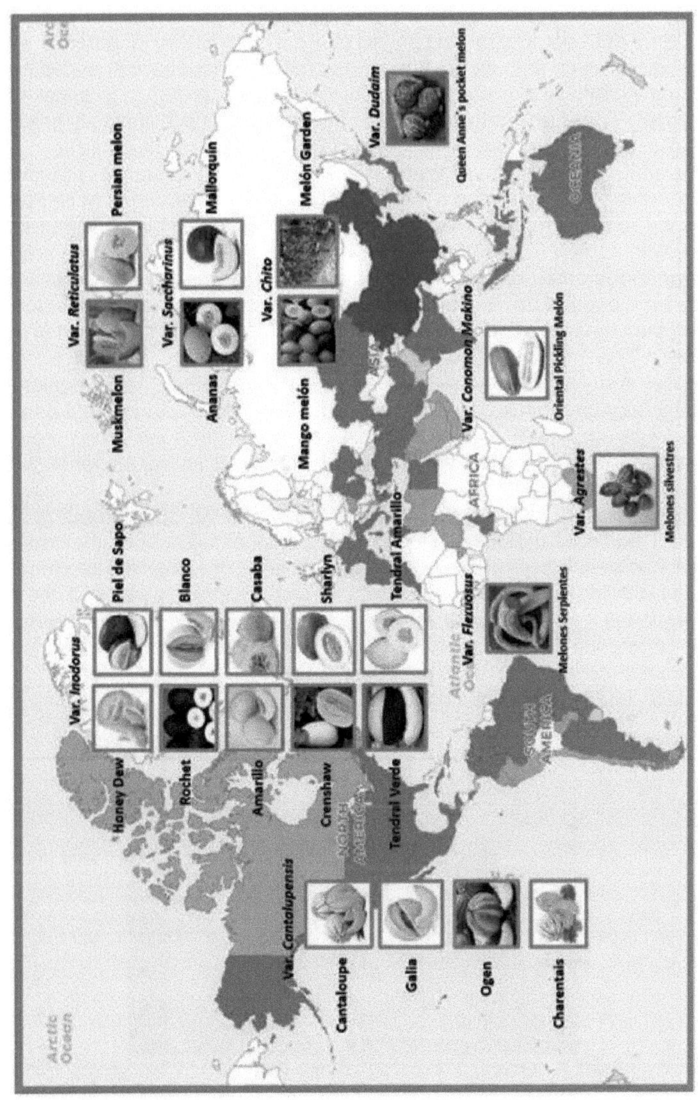

Types of melon varieties . Elaboration Veronica Garcia Mendoza

The melon that in our market and in that of the United States of America is known commercially like cantaloupe melon, whose name common most accepted today by the horticulturists is " *muskmelon* " (Reticulatus group), it is not the true "cantaloupe". The true "cantaloupe" (Cantalupensis group), is mainly produced in Europe for consumption own or in others places to export to said market [8].

With the internationalization of trade fruit and vegetable consumer markets They began to have the need to find a melon with enough conservation to be able to be transported to more remote areas . Initially it was resolved he problem marketing varieties with natural transportability such as melons Tendral , Yellows and Honey Dew even though they are not the top in quality organoleptic , even so^ the shelf life was not enough to reach more distant markets . The melons toad skin although they have enough natural transportability , its aspect external is not very attractive . The melons Charentais types began to have a good acceptance followed by the Galia type melon from genetics Israelite . consumption habits constantly they change and to satisfy are quality needs both internal as external and that also can reach regions far from where the produce , long melons are required life as he type harper they have characteristics improved genetics [26].

Hybrids

Melon plants have supported the selection effort and the plant breeding human . Methods Traditional melon cultivation has led to considerable improvement of varieties , however it is relatively slow and limited . it's possible produce Melon hyphorids with intraspecific variables among type genotypes wild and commercial melon varieties , with the goal of transferring some traits genetics particulars , such as resistance to fungi , bacteria , viruses and insects ; or tolerance to factors environmental conditions , such as salinity , flooding , drought , and temperature high or low , to commercial melon varieties .

Melon development in cultivation under agriculture protected (Photography Pedro Cano Rios)

Harper type hybrids

Harper - type melon hyphorids represent a opportunity for the melon producers not only export to distant markets , but also introduce this type of melon the national market and change he planting and marketing paradigm for the prolonged life postharvest that characterizes

In a study in the Comarca Lagunera 27 ' to evaluate shelf life , variables such as fruit shape , diameter equatorial and polar, in the Harper - type melon phorids were not modified significantly due to weight loss of the fruit caused by Over time ; the quality was seen diminished by he effect of storage up to 30 days at 18° Centigrade , also the weight loss of the fruit cause a reduction in firmness and

solid totals (°Brix):
Weight loss rate factors at 10 , 20 and 30 days of useful life in hphorids Melon evaluated at 18 ° Centtgrade room temperature

	PI (g)	PP %	Pp(g)	Fv (g/h)
10 days in	shelf (2	40 hours)		
Cruiser F1	2044	9.78	200	0.83
Alaniz Gold	1859	7.21	134	0.56
Queen RZ	nineteen ninety five	6.27	125	0.52
King RZ	2165	7.53	163	0.68
20 days in	shelf (4	80 hours)		
Cruiser F1	1859	26.14	486	1.01
Alaniz Gold	2091	12.67	265	0.55
Queen RZ	2175	11.08	241	0.50
King RZ	2282	12.58	287	0.60
30 d^as en	anaquel (7 20 h)			
Cruiser F1				
Alaniz Gold	1707	23.73	405	0.56
Queen RZ	1994	20.66	412	0.57
King RZ	2035	21.43	436	0.61

PI: initial weight , PP %: percentage of weight loss , Pp: weight lost , Fv : speed factor of weight loss in grams per hour

I agree with you results obtained , the Queen RZ and King RZ (Rijk Zwaan®) Harper -type phorides are marketable after 30 days of harvest , time Enough to reach distant markets without refrigeration . These hphorids also they had top characteristics for fruit shape , greater firmness and concentration of solids solid totals in comparison with the most common melon in the region, the htorido Cruiser F1 (Harris Moran®) and the Alaniz Gold (Sakata®).

hot melons type Harper King RZ (left) Queen RZ (right) (Rijk Zwaan® seeds) showing he peduncle attached (Photography Veronica Garda Mendoza)

Currently the hybrids Harper type in the Comarca Lagunera are produced only for the export market , since he seed price is higher than the conventional cantaloupe type of melon , resulting in a production cost high , that they would only pay the most demanding markets .

Requirements climate

Temperature

This crop is typical of areas with climates warm-dry , although support some times more temperate climates , although not cold . Seed germination occurs when he floor achieves a temperature of 22 to 30 °C, during he development vegetative of the plant must take care that it exists a temperature atmospheric from 25 to 30 °C and for flowering from 20 to 25 °C; well in This last process , the temperatures very high They tend to generate a greater number of male flowers [7] .

Humidity

The melon plant needs quite water in he growth period and during the maturation of the fruits to obtain good yields and quality . At the beginning of plant development , humidity relative should be 65 to 75%, in flowering of 60 to 70% and in fruiting from 55 to 65% [7] . The melon plant requires 686 grams of water to produce one gram of dry matter [28] .

Brightness

Lighting is important , especially during the growth periods initial and flowering . Light deficiency will have an impact directly in the decrease in the number of fruits in the harvest , like this same intensity luminica will determine the final relationship of male and female flowers, observing that in periods Short light conditions - eight hours of photoperiod - favor the production of female or hermaphrodite flowers [7] .

Requirements edaphic

The melon plant gives better results when grown in a soil with the following Characteristics : rich , deep, fluffy , well aerated , well drained , quite consistent , forming clumps. Does not provide good results on a soil that is excessively acid , tolerating floors slightly calcareous ; the pH that favors it is found between 6 and 7. It is necessary have well - drained soils whose subject content organic is acceptable . Furthermore, it is important that the floors be deep , approximately 60 centimeters deep at least [11] .

Requirements hydronic

water requirements in he cycle are 5,000 to 7,500 m ^{3} by hectare with one sensitivity to medium drought to high. During the first stages of your development , the water use is very low , as you go in the growing season he water use increases , due to an increase in solar radiation and temperature . The presence of a stress doctor in any of the phases phenological , production decreases , the most critical stage is in he flowering period so you must avoid humidity deficiencies [6] . Picture following shows the various stages phenological what you go through he melon cultivation .

Stages phenology of reticulated melon [6]

Phenological stage	Weeks after the emergency
Emergency	0
Plant	1
Vegetative development	4
Start bloom male	5
Start bloom feminine	6
Fruit binding	7
Fruit growth	8
Maturation	9

Sowing and transplanting

Planting in he sky cultivation opened is directly and can be mechanical or manual , two to three are deposited seeds by hitting one depth of 2 to 3 centimeters . Planting melon in our pa^s is done all he anus . In the Comarca Lagunera it is distributed from February to the end of May, although some producers start in the first week of January , looking for the inclination of the sun, for crops early ; the afternoons from August 15 to September 5 but sowings from September 5 to 10 are more recommended under irrigation by drip , though have he disadvantage that they can be affected by frost late or early respectively [6] .

Padded

The planting method and density is defined by the producer according to the availability of machinery and equipment , as well as the water source . The options to choose are:

• Melon beds 3 meters wide with double row of plants with plastic in the channel. The distance between plants is 30 centimeters

• Melon beds 1.8 meters wide with row simple in the center with ribbon and padding . The distance between plants is 25 centimeters

• Melon beds 1.6 meters wide with row simple in the center with ribbon and padding . The distance between plants is 20 centimeters . This system It allows mechanize he cultivation , facilitating pest and disease control is not necessary accommodate guides and the harvest can be perform with the use of trailers

The smaller the width of the beds and the smaller distance between plants the yields are higher ; that is , by increasing population density he performance tends to increase ; in this meaning with the bed system at 1.6 meters research results they have shown that the yields are higher by at least 20%, compared to that of 1.8 meter beds [6] .

Conclusions

after the continent Asian he American continent is the one that contributes the most in the participation of melon production at the level world . China generates more than half of the production being in the first place in the classification ; Mexico is located in he eleventh place . In our pa^s he Melon cultivation is important about all in the Laguna Region for being a generator of jobs , income important for producers agricultural companies and suppliers of goods and services . exist many varieties available , that adapt and give results in the different regions where it is grown , in he The most important varieties are Cantalupensis , Reticulatus and Inodorus; of the latter practically everything is exported for being a long harper type life of anaquel . The most important hyphorids have exquisite aroma , color, texture and sweetness. incomparable .

Melon garden in full harvest in Ceballos, Durango (Photography Veronica Garda Mendoza)

References

1.   Bisognin DA. Origin and evolution of cultivated cucurbits. Rural Science , Santa Maria.

2002;32(5):715- 23.

2.   Rosello JiO . Melons , a gift of summer . The fertility of the land. 2010;25:10 -4.

3.   Ayala-Aponte A, Cadena-G MI. The influence of osmotic pretreatments on melon (*Cucumis melo* L.) quality during frozen storage. DYNA. 2014;81(186):81-6.

4.   Nunez-Palenius HG, Gomez-Lim M, Ochoa-Alejo N, Grumet R, Lester G, Cantliffe DJ. Melon fruits: genetic diversity, physiology, and biotechnology features. Crit Rev Biotechnol. 2008;28(1):13-55.

5.   Rodriguez- Nodals AA, Sanchez-Perez P. Fruit tree species cultivated in Cuba in Urban Agriculture. 3rd Edition . Havana. 2005 [Online] https://www.ecured.cu/Mel%C3%B3n (Consulted 09/18/20).

6.   Chew MYI, Reyes JI, Espinoza AJdJ , Ramirez DM, Pastor LFJ, Figueroa VU, et al. Guide for melon production in the Lagunera Region . INIFAP-Comarca Lagunera . Mexico. 2010;1:17 p.

7.   CONABIO. National Commission for Knowledge and Use of Biodiversity . Organization Information System Alive Modified (SIOVM). GEF-CIBIOGEM Biosecurity Project . Mexico, DF 2006:1-28. [In line]. http://www.conabio.gob.mx/knowledge/biosecurity/doctos/consulta_SIOVM.html(Consulta 09/16/20).

8.   Fornaris GJ. Technological set for the production of "Cantaloupe" and "Honeydew" melons. University of Puerto Rico. Mayaguez University Campus. College of Sciences Agricultural . Agricultural Experimental Station 2001;1(1):1-5.

9.   Cano RP, Gonzalez VH. Effect of distance between beds on he growth , development and quality of fruits and production of melon *Cucumis melo* L. CELALA-INIFAP-SAGARPA. Matamoros, Coahuila, Mexico. Research Report .2002:342-5

10.   Fita A, Nuez F, Pico B. Adaptation of the root system of melon (*Cucumis melo* L.) against deficiency in match . Agricultural Vergel. 2011;1(1):151-4.

11.   Cano RP, Espinoza AJdJ . El Melon : Production and Marketing Technologies . Generalities of your production . CELALA-CIRNOC-INIFAP, Mexico. 2002;4(1):1-18.

12.   Monge-Perez JE, Loria-Coto M. Melon production in greenhouse : comparison agronomic between melon types . Postgraduate and Society. 2017;15(2):79-100.

13.   Saltveit ME. Volume 4: Mangosteen to White Sapote. *In* : Yahia E.(ed.) Postharvest biology and technology of tropical and subtropical fruits. Woodhead Publishing. 2011;536p.

14.   Yanty NAM, Lai OM, Osman A, Long K, Ghazali HM. Physicochemical properties of *Cucumis melo* var. Inodorus (Honeydew melon) seed and seed oil. J Food Lipids. 2008;15(1):42-55.

15.   De Melo MLS, Narain N, Bora PS. Characterization of some nutritional constituents of melon (*Cucumis melo* hybrid AF-522) seeds. Food Chem 2000;68:411 -4.

16.   De Mello MLS, Bora PS, Narain N. Fatty and amino acids composition of melon (*Cucumis melo* var. saccharinus) seeds. J. Food Comp Anal. 2001;14:69 -74.

17.   Exur Ltd. Directory of nutritional composition of melon. 2015. [Online] https://www.dietas.net/tablas-   y-calculadoras/tabla-de-compuesto-nutricional-de-los-alimentos/frutas/frutas-frescas/melon.html (Consultation 2/09 /twenty).

18.   FAOSTAT . Food and Agriculture Organization of the United Nations. 2020. [On line] http://www.fao.org/faostat/es/#data/QC/visualize (Consulted: 08/25/20).

19.   Monge-Perez JE. Costa Rican melon (*Cucumis melo*) production and exports. Technology on going. 2014;27(1):93-103.

20.   Lamez D, Krarup C. Characterization in pre and postharvest of two cultivars of reticulated melon of the type Eastern (*Cucumis melo* Cantalupensis Group). science Agrar Research . 2008;35(1):59-66.

21.   SAGARPA-INCA Rural. Secretary of Agriculture, Livestock , Rural Development, Fisheries and Food - National Institute for the Development of Capabilities of the Rural Sector AC System governing plan national product   melon2012.[Online] https://www.google.com/url?sa=t&rct=j&q=&esrc=s&source=web&cd=&cad=rja&uact=8&ved=2ahUKEwjGxoa-k43sAhUBSq0KHZjzDcAQFjAAegQIBhAB&url=http%3A%2F%2Fdev.pue.itesm.mx%2Fsagarpa% 2Fnational%2F EXP_CNSP_MELON%2FPLAN%2520RECTOR%2520WHAT%2520CONTAINS%2520PROGRAM%2520DE%252 0 WORK%25202012%2FPR_CNSP_%2520MELON_%25202012.pdf&usg=AOvVaw31Us_0BY3GIqHFCv DpoHu9 (Consultation 09/16/20).

22.   SIAP-SIACON-SADER. Information service Agri-food and Fisheries. Information system Agri-food Consultation. Secretary of Agriculture and Rural Development. [Online] https://www.google.com/url?sa=t&rct=j&q=&esrc=s&source=web&cd=&cad=rja&uact=8&ved=2ahUKEwibjKb_ko3s A   hUOPK0KHU25D1IQFjAAegQIBBAB&url=https%3A%2F%2Fwww.gob.mx%2Fsiap   %2Factions-and-programs%2Fagricultural-production-33119&usg=AOvVaw1glm7RJPKJwq9s7sYRa2pB (Consulted 09/16/20).

23.   Hernandez- Martmez J, Garda-Salazar JA, Mora-Flores JS, Garda-Mata R, Valdivia-Alcala R, Portillo-Vazquez M. Effects of the elimination of tariffs on melon (*Cucumis melo* L.) exports from Mexico to the USA . Agroscience . 2006;40(3):395-407.

24.   Rairurez -Barraza BA, Garda-Salazar JA, Mora-Flores JS. Production of melon and sand^ a in the Lagunera Region : a planning study to reduce price volatility . Ergo-sum science . 2015;22(1):45-53.

25.   Horvath L, Gyulai G, Szabo Z, Lagler R, Toth Z, Heszky L. Morphological diversity in yellow melon (*Cucumis melo*); one medieval tfpus species reconstruction . Agricultural science Events .

2007;(27):84-90. Hungarian .

26.   Soriano-Vicente B. Morphogenesis *in* vitro and plant breeding transgenic melon (*Cucumis melo* L.). Thesis. University of Almeria-Higher School of Engineering-Department of Plant Biology and Ecology . Almeria, Spain. 2012:1-93. [Online]   https://core.ac.uk/download/pdf/143455655.pdf (Accessed 09/30/21)

27.   Garda-Mendoza V, Cano-Rfos P, Reyes-Carrillo JL. Harper-type melon hybrids have higher quality and

longer post-harvest life than commercial hybrids. Rev Chapingo Ser Hortic . 2019; 25(3), 185-97.

28.   Covers PR. Operation manual agronomic for melon cultivation . *Cucumis melo L.* Agricultural Development Institute - Research Institute Agricultural . Department of agriculture. Government of Chile. 2017;1:92 p .

"In the village, there is no bad melon neither women ugly "

anonymous

8. Introduction and removal of colonies from the crop
Pedro Cano-Riosy Jose Luis Reyes-Carrillo Introduction

In the Mexican Republic , melon is one of the most important vegetables . The surface busy by this level cultivation national in he year 2019 was 19,838 hectares with a production of 627,135 tons , equivalent to an average of 31.6 tons by hectare . The most important states by his surface planted are Coahuila, Sonora, Guerrero, Michoacan, Durango, Chihuahua, Colima, Oaxaca, Jalisco and Guanajuato [1] .

In the Laguna Region the melon is one of the more remunerative crops and more labor occupies during he cycle spring- summer agricultural . Is by Consequently, the vegetable of greatest social and economic importance , in this agricultural area . Only in he In 2019, 5,118 hectares of melon were planted , with a yield regional average approximately 34.2 tons per hectare [2] ; In addition , La Laguna is the most important region in melon cultivation at level national with 25 per percent of the country 's surface , being The municipalities of San Pedro, Viesca and Matamoros are the largest [3] .

Besides issues inherent to the cultivation of melon, in the Lagunera region stands out he little or none use of agents pollinators to ensure a good " tie " of the harvest [4] . This reflects a unlogical carelessness about the physiology of the plant, in particular with respect to its pollination . Melon flowers are attractive as food source for bees [5] and these they can gather he pollen from reproductive structures [6] . The Mbridos Current melon plants have staminate flowers and hermaphrodite flowers. on the same floor, offering a nutritious reward for visitors in exchange for these disperse pollen grains taking them to other flowers to be fertilized [7] :

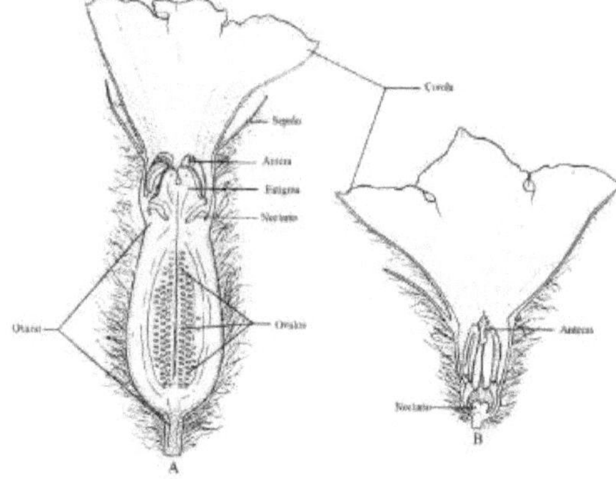

Hermaphrodite flower (left) and flower male (right) present simultaneously in a melon plant. Drawing taken from McGregor [7] . Dissection and photographs Olga Araceli Zapata Ramos.

A study done in the Comarca Lagunera , revealed that by isolating the hermaphrodite melon flowers with mesh hoods to avoid the visits of the pollinating insects the fruits did not " bind ". This was because although self - fertilization is possible in flowers, this does not happen since the Melon pollen is heavy and sticky and is very it is difficult for the grains to travel alone to it part female of the same or another flower . So it is essential that it be transferred by insects [8] .
The number of visits to the flower by part of the pollinators has effect straight about he yield and quality of the fruit , therefore , the greater the number of visits the greater the number of seeds , an indicator of quality . like the seeds produce fruit growth hormones , at least they must be develop 400 seeds by fruit so that the melon has commercial acceptance [7] .

reticulated melon game in half showing the seeds (Photography Jose Luis Reyes Carrillo)
To achieve a good pollination must be cover four basic points :
1. Stop pesticide applications especially during he d^a , to avoid give us the bees
2. Place the hives in the growing area at the beginning of flowering male , or slightly before flowering female or hermaphrodite . It is not recommended place them very anticipation , since they will look for others crops to maintain and when needed in the melon will be difficult return them . This happens because melon

flowers offer pollen in abundance but not so much nectar. Thus can It may happen , although not always , that the bees become distracted looking for nectar others places visiting except the melon flowers

3. Place the drawers in direction favorable to air currents , so that it helps them in he flight

4. Place the hives in he side opposite to the water supply source , so that between the two there is he cultivation , to force them to fly over it constantly [9].

The above are simply the generalities of management efficient to pollinate he melon cultivation . For one more information We recommend reading others treaties much more complete published by various authors [10-12]. However, it is necessary clarify that even in those jobs not required how much is lost in production and quality of fruit by a introduction afternoon of the bees neither when they are due remove colonies from growing areas .

With the purpose of giving answer To answer both questions , two studies were carried out in a melon crop with bees meKferas in periods different pollination corresponding to two consecutive years [13]. The results are shown below .

Studies

Are investigations were carried out in the INIFAP La Laguna Experimental Field, in Matamoros, Coahuila, within the Lagunera Region . The average rainfall in This area is 235 miHmeters per year . It is located one height about he sea level of 1,139 meters and presents a average annual temperature of 18.6°C [14].

Location of the studies

The experiments were set during he spring summer cycle ; at the field level jobs made were the equestrian preparation of the land as fallow , cross tracking , leveling and with planting beds 1.80 meters wide, sowing a row of plants in the center of the bed , spaced every 20 centimeters according to what is recommended for the Lagunera Region [15]. The hyphorids used for planting They were Gold Rush for the first year and Cruiser for the year following ; the plantings were carried out April 18 and April 26 for the first and second anus , respectively . The crop was left to free growth vegetative in the two years of cultivation . The grooves were mulched with black plastic and fertigation by means of tape . For him insect supply pollinators settled five bee hives Jumbo size with some queens commercial new and standardized to about 24,000 bees workers , considered by diverse sources as enough to make a correct pollination [16-19].

The studies were developed during he spring summer cycle . Between the jobs field preparatory work , the equestrian preparation of the land was carried out as fallow , the cross tracking , leveling the surface with planting beds 1.80 meters wide. It was planted a row of plants in the center of the bed , spaced every 20 centimeters in accordance with what is recommended for the Lagunera Region [15]. The hyphorids used for planting They were Gold Rush for the first year and Cruiser for the second . The plantings were done April 18 and April 26 for the first and second anus , respectively . The crop was left to free growth vegetative in the two years of cultivation .

The grooves were padded with black plastic and applique fertigation by means of tape . For him insect supply pollinators settled five hives Jumbo size , with queens fertilized commercial standardized to about 20,000 bees workers . The different isolation treatments were covered with Agribon [®] which is a Non - woven fabric cover , ultralight and resistant to environmental exposure that allows the passage of light, water and air to isolate the flowers from the action of pollinators . Previously applied the insecticides Mitac 20® CE and Endosulfan 35® at 1.5 liters by hectare in each of the treatments for mosquito control silver leaf white and others insects plague .

Melon cultivation beds isolated through Agribon ® and hives pollinators (Photographs Jose Luis Reyes Carrillo)

Studies
The study of pollination periods for the first year was the following :
1. Pollination from the first flowering week (without covering)
2. Pollination from the second flowering week
3. Pollination from the third flowering week
4. Pollination from the fourth flowering week
5. Pollination from the fifth flowering week
6. Pollination the first week and covered with Agribon ® in the following weeks
7. Pollination until the second week and covered with Agribon ® in the following weeks
8. Pollination until the third week and covered with Agribon ® in the following weeks
9. Pollination up to the fourth week and covered with Agribon ® in the following weeks .

For him study of the second year a tenth treatment was added , the which remained covered with Agribon ® all he cycle .

The plot for data collection was a bed melonera 10 meters long by 1.8 meters wide for the first year , while the experiment of the second year had a Useful plot eight meters long by 1.8 meters wide.

Characteristics evaluated
Within each plot were evaluated the returns according to the quality of the fruit : the export one , which are those with less than 5% external staining ; the consumer national , with a stain in the shell greater than 5% and less than 15% or with a certain deformity ; that of lag or waste , which includes the fruits with spots or damage greater than 15%. It was measured he size of the fruits in a size scale conventional 9, 12, 15, 18, 23 or 30; these are known as packaging categories . These numbers represent he number of melons that fit on the fence melonera packaging standard . For the measurement, it is used a wooden board with holes by where pass or not fruits . performance commercial was calculated adding up the returns export and national . During the second year I only estimate he performance national .

Determination of the size and weight of the melon fruit (Photography Pedro Cano Rios)

Analysis of data . The data obtained were analyzed to detect differences and trends between treatments [20] .

Results

First year experiment

The delay in he start of pollination affected the quality of the fruits , especially the quality export type , given that it was reduced gradually from 41.1%, which was registered in the first week of flowering , up to 7.9% in the fifth week . were not observed differences in he percentage of performance between the different melon classes in the first and second flowering week how to appreciate in he following chart :

Effect of the start of pollination by bees on the quality of the melon harvest the first year

Start of pollination (flowering week)	% of performance by category		
	Fall behind	National	Export
1	25.7	33.2	41.1
2	24.9	30.0	45.1
3	35.3	26.0	38.7
4	38.0	33.1	28.9
5	63.7	28.3	7.9

The delay in pollination cause also a delay in the harvest , observing a marked difference When pollination began in the third week in forward ; by example in the weeks first and second harvested 38.9 and 31.6 %, respectively , for the weeks third , fourth and fifth pollination , they have been harvested only 7.5 , 0.8 and 0%, respectively :

Effect of the start of pollination by bees in he delay of the melon harvest the first year

Start of pollination (flowering week)	yield percentage by harvest date					
	Oct. 7	Oct. 11	Oct. 13	Oct. 18	Oct. 25	Nov. 1
1	9.7	30.2	38.9	68.3	88.6	100
2	12.8	23.6	31.6	61.2	79.7	100
3	0.8	1.9	7.5	34.3	75.8	100
4	0.4	0.8	0.8	12.0	57.2	100

5	0	0	0	0	0	100

The number and weight of fruits of larger size categories as he number 9, 12 and 15 fruits by grate , they were reduced in a proportion even greater when pollination I started later . The above is because to obtain a fruit melon commercial requires several hundreds of pollen grains are deposited in he stigma of each flower hermaphrodite and to achieve the above, according to McGregor [7] each flower hermaphrodite should be visited between 10 and 15 times during he d^ a when it opened While in the soursop, for For example , at least four beetles are required pollinators by flower to generate a fruit in a regular way [21] .

If the pollination of the melon is deficient [10] , we obtain fruits with less seeds and in consequence , deformed or minor size . Similar results were found by researchers from the Department of Agriculture of the United States [22] in Weslaco, Texas with treatments of 0, 6 and 12 days of delay in pollination of the melons Cruiser and Explorer vehicles that produced smaller fruits when pollination was delayed 12 days , however, they did not find effects negative when pollination was delayed for six days . In it cucumber cultivation in the flowers that produced fruit was found a association between a greater number of visits by bees and the greater time accumulated of the visits with the highest commercial quality [23] . Fruit weight and characteristics qualitative related to the way he diameter equatorial and longitudinal too are related to the visits of the insects to the flower , given that the flowers with the greatest number of visits and the longest time accumulated visits , they had also the higher yields [23] .

Melon packaging size in the typical fence Melonera (Photography Pedro Cano Rios)

Performance first and second commercial year

The data collected during the two years for him performance commercial they threw nails differences for pollination periods . Considering that it was observed a association between beginnings of pollination and performance , with data of the treatments initials (1, 2, 3, 4, 5) was found a answer significant among said treatments and performance commercial , that is, the earlier pollination , the higher the yield , and the longer the delay , the lower the yield . performance , as illustrated by the following graph .

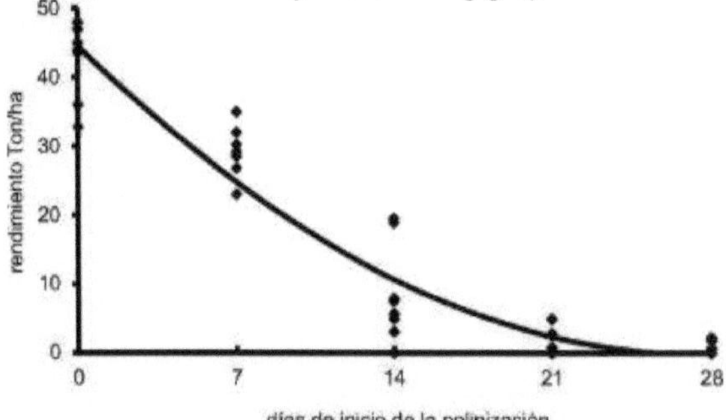

Relationship between start of pollination and performance melon commercial in the experiments from the two years of study

The graph indicates that starting pollination When the first hermaphrodite flowers appear , that is , 0 days late , there will be a yield 44.5 ton commercial by hectare . The above implies that 22.2 tons / ha would be lost if pollination begins 7 days after flowering begins . hermaphrodite Since more than one is required pollinator visit to the flower , producers need provide the pollinators suitable in efficiency and in number to reach he ideal load level in he moment opportune , because under certain conditions the advantage number of bees is reduced due to the presence of other flowers that attract them and keep them away from the melon crop [25] .

Bee visiting the flower male melon (Photography Jose Luis Reyes Carrillo)

Removal of hives

To answer the question of how much time are due leave the bees in the farmland , were used the treatments 1, 6, 7, 8, 9 and 10 , as seen on the graph following :

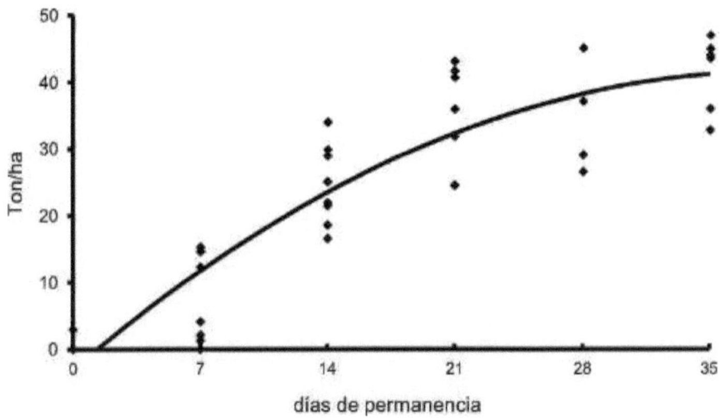

Relationship between days of permanence of bee hives in he cultivation and performance melon commercial in the experiments from both years

The graph indicates that after 28 days of stay of the bees in he cultivation has a yield of 38.3 tons by hectare , while after 35 days there is a yield of 41.3 tons by hectare . This means that between both indices there was no a difference important , that is , that since From a performance point of view , bees can be withdraw around 28 days after having the first hermaphrodite flowers appeared without affecting he number of melons " tied " and by hence , production .

Harvested melons showing the differences in quality in the different categories (Pedro Cano Rios Photography)

The effect in he binding of the fruit and therefore , in he performance is well known in fruit trees and crops that are visited for the bees during flowering [7.23] and crops foragers pollinated to produce seed [26] . For another side , in the pollination of the radish by For example , it was found that the amount of pollen hauled for the bees foragers related positively with the number of visits to the flower [27] .

Conclusions

The results of these two years of study allow conclude that the delay in he initiation of pollination causes a significant delay in the harvest , which has an effect negative in the quality of the fruit , especially in the best quality , which is reflected in a significant decrement in the weight, the number of fruits , the size and the melon yield , there is a relationship between start of pollination and the performance commercial . They can be remove the hives 28 days after having the first hermaphrodite flowers appeared since it is enough time <u>to achieve he maximum</u> performance <u>of</u> melon cultivation .

Open melon flowers waiting for the visit of the insects (Photography Olga Aracely Zapata Ramos)

References

1. Information service Agri-food and Fisheries (SIAP). Melon statistics in Mexico. 2019. [Online] https://blogagricultura.com/estadisticas-melon-mexico/ (Consulted 01/09/20).
2. Secretary of Agriculture and Rural Development (SADER). Agricultural Summary of the Laguna Region in 2019. 2020; *In:* El Siglo de Torreon special edition January 1, 2020.
3. Espinoza-Arellano JdJ , Orona-Castillo I, Guerrero-Ramos LA, Molina-Morejon VM, Ramirez-Quiroga EC. Analysis of financing, marketing and profitability of melon with a "sowing by stages" approach in the Comarca Lagunera region of Coahuila state, Mexico. UAT Science ;2019;13(3): 71-82.
4. Ayala , J. Identification of melon (*Cucumis melo* L.) production systems in the Lagunera and Parras de la Fuente Coahuila region . [Master 's Thesis] Torreon, Coahuila, Mexico. autonomous University Antonio Narro Agrarian , Laguna Unit. 1997; 203p.
5. Delaplane, KS Bee pollination of Georgia. Crop plants. University of Georgia. College of Agricultural and Environmental Science. Cooperative Extension Service Bulletin 1994;1106:37p.
6. Labandeira, CC How old is the flower and the fly? Science 1998;280(5360): 57-9.
7. McGregor SE. 1976. Insect pollination of cultivated crop plants. USDA, Agricutural. Handbook U.S. Government Printing office, Washington, D.C., U.S.A. 1976; 411p.
8. Reyes-Carrillo JL, Valdez-Perezgasga MT, Villa-Carrera, DM. Pollination by bees (*Apis mellifera* L) in he melon cultivation (*Cucumis melo* L.) in the Lagunera region . Association Latin American Sciences Agricultural . 1982;17(1): 17-28.
9. Sabori-Palma R, Grageda-Grageda J, Chavez-Cajigas JM, Fu-Castillo AA. Guide for the production of cucurbits on the Coast of Hermosillo, National Research Institute Forestry , Agricultural and Livestock (INIFAP) - Northwest Regional Research Center (CIRNO) Costa de Hermosillo Experimental Field (CECH). User information Technical second edition. 2004;16: 139p.
10. DeLaplane KS, Mayer DF. Crop pollination by bees. University Press Cambridge, UK 2000; 352p.
11. Reyes-Carrillo JL, Cano-Rtos P. Pollination Manual Apfcola . National Program for the Control of the African Bee-Inter-American Institute for Agricultural Cooperation. Manual. 2002;7: 52p.
12. Reyes-Carrillo JL, Eischen FA, Cano-Rtos P, Nava-Camberos U. Pollen collection and honey bee forager distribution in the cantaloupe. Acta Zool Mex (ns .) 2007;23:(1): 29-36
13. Reyes-Carrillo JL, Cano-Rtos P, Nava-Camberos U. Periodo optimo de polinizacion del melon con abejas meKferas (*Apis mellifera* L.). Agric Tec Mex. 2009;35(4): 371-378
14. Schmidt RH. The arid zones of Mexico: climatic extremes and conceptualization of the Sonoran Desert. J. Arid Environ. 1989;16:241-56.
15. Cano-Rtos P. 1992. New system melon grower for the Lagunera County Journal of Vegetables , Fruits and Flowers. Dec 1992: 19-24.
16. Atkins EL, Mussen E, Thorp R. Honey bee pollination of cantaloupe, cucumber and watermelon. Division of Agricultural Sciences. University of California. 1979;Leaflet 2253 .
17. DeLaplane KS, Mayer DF. Crop pollination by bees. University Press Cambridge, UK 2000; 352p.
18. Eischen FA, Underwood BA. Cantaloupe pollination trials in the lower Rio Grande Valley. Am. Bee J. 1991;131(12):775.
19. Hodges L, Baxendale YF. Bee pollination of cucurbit crops. University of Nebraska-Lincoln. Cooperative Extension. Institute of Agriculture and Natural Resources.1995;Bulletin NF91-5D: 2p.
20. Steel RGD, Torrie JH. Principles and procedures of statistics. McGraw-Hill BookCompany, New York, USA. 1960; 481p.
21. Podoler H, Galon I, Gazit S.The effect of *Atemoya* flowers on their pollinators: *Nitidulid* beetles. ActaEcologica 1985;6(3) 251-58.
22. Eischen FA, Underwood BA, Collins AM. The effect of delaying pollination on cantaloupe

production. J Apic Res. 1994;33(3):180-4.

23.   Gingras D, Gingras J, De Oliveira D. Visits of honeybees (Hymenoptera: Apidae) and their effects on cucumber yields in the field. HortEntomol. 1999;92(2): 435-8.

24.   Klein AM, Steffan-Dewenter I, Tscharntke T. 2003. Bee pollination and fruit set of *Coffea arabica* and *C. canephora* (Rubiaceae). Am J Bot. 90(1): 153-7.

25.   Dogterom MH, Winston ML, Mukai A. Effect of pollen load size and source (self, outcross) on seed and fruit production in highbush blueberry cv. " Bluecrop " (*Vaccinium corimbosum* , Ericaseae). Am J Bot. 2000;87(11):1584-91.

26.   Levin, MD Distribution patterns of young and experienced honey bees foraging on alfalfa. J Econ Entomol . 1959;52:969–71.

27.   Rush S, Conner J, Jennetten , P. The effects of natural variation in pollinator visitation on rates of pollen removal in wild radish, *Raphanus raphanistrum* (Brassicaceae) Am J Bot. 1995;82(12):1522-6.

" Success is where preparation and opportunity meet " Bobby Unser

9. Distance from the hives to the melon crop

Jose Luis Reyes-Carrillo and Pedro Cano -Rios

Introduction

The success in the pollination of a plant is certain to a large degree by three factors : the number of pollinators that visit the plant, the number of flowers it visits each pollinator during his plant visit and pollinator effectiveness in the transfer appropriate pollen stamens to the pistil of a receptive flower [1] . The plants introduced by the humans to an ecosystem in where they would not exist previously they can be especially in disadvantage in this respect , since they have not evolved together with the local pollinators . Can give he case that some local pollinator sits attracted for the flowers but it does not have the characteristics anatomical to access them and therefore it will not be a pollinator cash . Likewise , you can to have pollinators that chance have the characteristics anatomical appropriate to access the flowers and to pollinate them , but that are not attracted with success for them [2] . The visitors floral they continue search strategies certain by preferences innate of certain traits and signs distinctive features of the flowers or preferences learned Modifiable depending on experience and rewards [3-5] . The flowers that require pollination crusade to produce fruits depend on a greater number of visits and a longer duration cumulative of these visits by pollinators [6] .

The bees meKferas have used as pollinators of the melon crop with a lot success because they comply amply with the three factors that we mentioned , that is , there are many bees that visit the flowers , each bee visit many flowers and they are also very efficient as pollinators . To get the greatest benefit of these insects , the hives are placed in quantity and location you specify in the growing areas during he flowering period . However there is a problem inherent to are bees that started when bees arrived in Mexico Africanized . The problem lies in which for his unpredictable aggressiveness represent a risk to day laborers , particularly during irrigation and weed control tasks . In 1990, with the arrival of the bees Africanized to Texas began a new period of colonization [7] from which They migrated to northern Mexico. The bees meKferas africanized They arrived from South America through the state of Chiapas [8] . A method to reduce he impact in pollination for the bees Africanized , is to make the crops aim are more attractive to bees hive areas that are available [9] and place the colonies close to the crop since it is well known that the pollinator choose flowers according to reward and energy expenditure [10-12] and more recently it is known that also he attractive related to aromas , nectar production , pollen and flower color [13,14] .

For this reason , in the Comarca Lagunera , it is not allowed locate the hives within the melon crop , but on the periphery . The producers they request the beekeepers who apiaries installed with the goal of pollination these crops are placed outside the garden to avoid bite problems . Since melon flowers depend on these bees to be pollinated and to achieve the highest performances possible , this chapter illustrates the way in which the distance of bee colonies from melon plants is related to he yield and quality of the fruit .

worker bee visiting a flower male melon (Samuel Atahualpa RamirezMacias Photography)

The study

This research was carried out during spring and summer in a crop commercial Imbrido Cruiser melon of six hectares in the PP Las Cruces, municipality of Viesca, in he state of Coahuila. The site corresponds to the Comarca Lagunera region , which has a semi-arid climate [15] . The crop rectangle measured at 105 meters wide and 571.42 meters long. The grooves They ran the width of the crop , that is , they measured 105 meters. To pollinate he cultivation were used three hives by hectare , Jumbo size , each one with one queen new and one population of approximately 24,000 bees workers . The 18 hives were distributed evenly along one of the two longest sides (571.4 m) of the crop . Four rows were selected at random. complete (105 meters each) and along them were marked Kneas (transects) 10 meters long whose center was located 25, 50, 75 and 100 meters from the nearest hive . In each transect , the distance from the crown (center) of each plant to its " trunk " melon was measured as calls he farmer at melons closest to the crown, it was counted he number of melons , and they were measured he longitudinal diameter , circumference equatorial and the individual weight of each fruit .

Melon plants in a 10 - meter transect where the reference stake is observed (Photographs Jose Luis Reyes Carrillo)

Field data were examined assuming the distance to the hive as the variable of importance to analyze to compare the fruit averages [16] .

Results

The objective of this study was determine the relationship between the distance of the hives from the melon crop and the fruit size . The distance between the melon and the crown was measured by be a related variable positively with the size and weight of the fruit . This is because the proximity of the fruit to the main stem gives greater accessibility to the items nutritious . This distance between main stem of the plant and the fruit has a relationship direct with him start time of flower pollination , since a pollination delayed it would turn out in a greater distance between the two. However, the results of this work they did not reveal none relationship or difference between the distance of the hive and the plants , and the distance between the crown of the plant and the fruit trunker . The next board sample the results .

Effect of distance between hives and plants in the average distance between fruit trunk and crown, with respect to the size of the melon.

distance	Distance to crown (cm)	circumference equatorial (cm)	length (cm)
25m	25.4	49.1	20.9
50 m	29.1	39.5	20.1
75 m	38.9	44.5	23.0
100 m	28.7	41.0	21.0

The size of the fruit did not suffer none effect regarding the distances from the hives to the plants , since no differences in the circumference neither in he longitudinal diameter , that is , it was not found a relationship between the distances that exist on one side between the crown of the plant and and the first fruit and the distance between the hives and the plants . The circumference equatorial and the longitudinal diameter related to the distance of the hives were similar and were not found relationships between distances and size variables ; this means that the Melon size was not related to the distance from the apiary to the plant.

Measures related to the performance as he number average fruit per plant and individual melon weight are shown In the table following :

Effect of distance from the apiary in he average number of melons per plant and weight of the melon

distance	melons /plant	Melon weight (kg)
25 m	1.4 a	1.41
50 m	1.5 a	1.42
75 m	2.0 a	1.58
100 m	1.7 a	1.72

Melon plants showing the melons logs in development (Olga Araceli Zapata Ramos)

It was not observed trend some to have more fruits per plant at shorter distances from the bee colonies since the average fruit in the different apiary distances were equals . The average weights individual melons among the different apiary distances were also equals .

As it was mentioned already Previously , the main objective in this study was determine the influence of the distance of the hives pollinators in the production and quality of melon. We already talked about the variants of the study , now lack describe the relationship of results with bees .

The bees meKferas frequently they visit sequentially the flowers of a single species , although fly about others that also are available . Your behavior always tour in function of the reward , although not always by quantity or quality . This " floral constancy " that is characteristic of these insects in relationship to your feeding , has been object of many studies . Even there are also with others bees and others insects , like flies , beetles and butterflies [17-19] .

A bee meKfera can fly one considerable distance to collect nectar or pollen , in average 3.5 kilometer radius with respect to its hive and in cases extremes up to 13 km , but a time there , it will tend to work in a small area [12] . This would explain why there has not been observed a difference between the quantity and quality of melon fruit in different distances from the experiments , that is , that these distances were insignificant for the distances that bees are capable of traveling . Likewise , these results are seen reinforced by he fact that the grooves in the cultivation area studied , measured only 105 meters long.

Once it is known behavior bee food meKfera , this can be easily manipulated to obtain a high fidelity [20] . This can explain your feeding activities in the vicinity of the apiary at distances evaluated . At our experience , the beekeepers and producers They try to locate the hives near the melon crop to reduce he flight time , then spending more time in the search for food reduces the lifespan of the bee [21] .

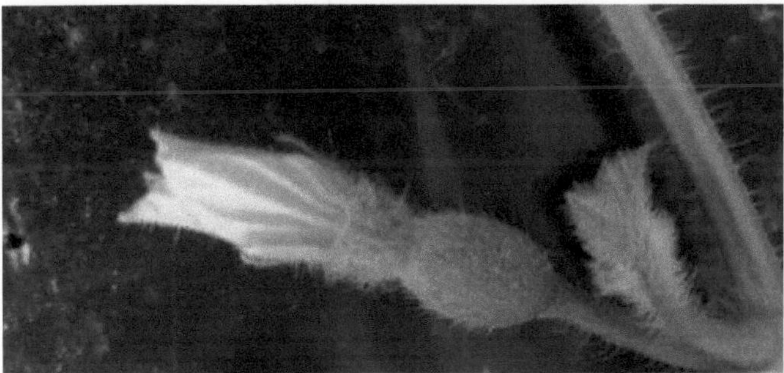

Hermaphrodite flower fertilized in the melon cultivation (Photograph Olga Araceli Zapata Ramos)

A fruit tie abundant of " trunneros " melons is always desirable , but even more so when it comes to a harvest early , since this means maturation stage concentrated in he time , superior quality of the fruit and consequent top prices in the market. The distance between the crown and the fruit is a extent related to the moment when pollination begins after the plant begins to flower , and it is a tool powerful for studying synchrony between plants in flower , pollination with bees and fertilization of the ovary [22] . In this work was observed a consistent pollination uniformity since there was not none difference between distances analyzed from he apiary .

Fruit weight and characteristics related to shape, such as he diameter equatorial and longitude , are related to the visit of flowers by part of the insects pollinators , since in Similar studies , the flowers they had the greatest number of visits and durations highest accumulated of those visits , too they had the higher yields [23], [24] . The above is seen strengthened by he fact that melon flowers are attractive as food source for bees [23,25] and these they can carry he pollen to structures reproductive of flowers [19,23,25].

In others cucurbits as in the cucumber, the pollination rate is correlated positively with the weight and maximum circumference , but not with the length of the fruit [6]. This pecoreo effect about he performance is well known in trees fruit trees and crops that are visited frequently for the bees during his flowering period [26] and widely acquaintance in crops foragers pollinated to produce seeds [12] ; since the bees meHferas they can provide he pollination service for many crops and plants ornamentals that compete with others insects per food [27] in this uniformity work in the distribution in the cultivation area of these insects in his visits to the flowers he was important to pollinate and influence in production and quality homogeneous melon.

Melon cultivation with fruits in development after pollination with bees meliferas (Photographs Jose Luis Reyes Carrillo)

Conclusions

Under the conditions of the study and according to the observations made during he field work , in this crop commercially , pollination was carried out in a uniform at distances studied since they were not found differences in performance variables evaluated , and therefore it can be conclude that, in the different distances from the hives to the melon crop , the bees pollinators were distributed uniformly and did not affect the characteristics of the fruit .

Frutos de melon a granel antes de la selection y empaque (Fotografia Jose Luis Reyes Carrillo)

Referencias

1.   Cresswell JE. The influence of nectar and pollen availability on pollen transfer by individual flower

of oilseed rape (*Brassica napus*) when pollinated by bumblebees (*Bombus lapidarius*). J Ecol. 1999;87(4):670-7.

2. Dudareva N, Pichersky E. Biochemical and molecular genetic aspects of floral scents. Plant Physiol. 2000;122:627-33.

3. Lunau K. Adaptive radiation and coevolution — pollination biology case studies. Org Divers Evol. 2004;4:207-24.

4. Pelz C, Gerber B, Menzel R. Odor and intensity as a determinant for olfactory conditioning in honeybees: roles in discrimination, overshadowing and memory consolidation. J Exp Biol. 1997(200):837—47.

5. Kearns CA, Inouye DW, Waser N. Endangered mutualism: The conservation of plant - pollinator interactions. Ann Rev Ecol Syst. 1998;29:83-106.

6. Gingras D, Gingras J, De Oliveira D. Visits of honeybees (Hymenoptera: Apidae) and their effects on cucumber yields in thef ield. Hortic Entomol. 1999;92(2):435-8.

7. Pinto AM, Johnston JS, Rubink WL, Coulson RN, Patton JC, Sheppard WS. Identification of Africanized honeybee (Hymenoptera: Apidae) mitochondrial DNA: validation of a rapid polymerase chain reaction-based assay. Ann Entomol Soc Am. 2003;96(5):679-84.

8. Fierro MM, Munoz MJ, Lopez A, Sumuano X, Salcedo H, Roblero G. Detection and control of the Africanized bee in coastal Chiapas, Mexico. Am Bee J. 1988;128(4):272-5.

9. Ambrose JT, Schultheis JR, Bambara SB, Mangum W. An evaluation of selected commercial attractants in the pollination of cucumbers and watermelons. Am Bee J. 1995;134:267-71.

10. Lee WR. The non random distribution of foraging bees between apiaries. J Econ Entomol. 1961;52:928-33.

11. Waser NM, Chittka L, Price MV, Williams NM, Ollerton J. Generalization in pollination systems, and why it matters. Ecology. 1996;77(4):1043-69.

12. Levin MD. Distribution patterns of young and experienced honey bees foraging on alfalfa. J Econ Entomol. 1959;52:969-71.

13. Varassini IG, Trigo JR, Sazima M. The role of nectar production, flower pigments and odour in the pollination of four species of *Passiflora* (Passifloraceae) in south-eastern Brazil. Bot J Linn Soc. 2001;136:139-52.

14. Briscoe A, Chittka D. The evolution of color vision in insects. Annu Rev Entomol. 2001;46:471-510.

15. Schmidt RH. The arid zones of Mexico: climatic extremes and conceptualization of theSonoran Desert. J Arid Environ. 1989;16:241-56.

16. Steel RGD, Torrie JH. Principles and procedures of statistics. McGraw-Hill Book Company, Inc. 1960;New York, USA:481p.

17. Gegear RJ, Laverty TM. How many flowers types can bumblebees work at the same time? Can J Zool. 1998;76:1358-65.

18. Gegear RJ, Laverty TM. The effect of variation among floral traits on the flower constancy of pollinators. *In:* Chittka L, Thomson J D editors Cognitive ecology of pollination: animal behavior and floral evolution. Cambridge University Press, Cambridge, UK. 2001;1-20.

19. Labandeira CC. How old is the flower and the fly? Science. 1998;280(5360):57-9.

20. Meller VH, Davis RL. Biochemistry of insect learning: lessons from bees and flies. Insect Biochem Molec Biol. 1996;26(4):327-35.

21. Hrassnigg N, Crailsheim K. The influence of brood on the pollen consumption of worker bees (*Apis mellifera* L.). J Insect Physiol. 1998;44:393-404.

22. Peet M. Sustainable practices for vegetable production in the South. Muskmelon. Focus Publishing/R. Pullins Co. Newburyport, MA. USA. 1996;174p.

23. McGregor SE. Insect pollination of cultivated crop plants. Agriculture Handbook No 496 United States Department of Agriculture. Washington, D.C.1976;411p.

24. Richards AJ. Does low biodiversity resulting from modern agricultural practice affect crop pollination and yield? Ann Bot. 2001;88(2):165-72.

25. Eischen F, Underwood BA, Collins A. The effect of delaying pollination on cantaloupe production. J Apic Res. 1994;33(3):180-84.

26. Klein AM, Stefaffan-Dewenter I, Tscharntke T. Bee pollination and fruit set of *Coffea arabica* and *C. canephora* (Rubiaceae). Am J Botany. 2003;90(1):153-57.

27. Comba L, Corbet SA, Barron A, Bird A, Collinge S, Miyazaki N, et al. Garden flowers: insect visits and the floral reward of horticulturally-modified variants. Ann Bot. 1999;83(1):73-86.

"In time of melons , short the sermons "

Anonymous

10. Behavior of bees during pollination in melon crops

Jose Luis Reyes-Carrillo and Pedro Cano-Rfos

Introduction

For many crops , the performance and quality would be reduced considerably without the pollination of honey bees [1] . The poor quality of the fruit It is generally attributed to pollination problems , such as They may be a low number of bees or the inefficiency of the pollinators within agroecosystems [2,3] . The amount of pollen that a hive have stored , in he case of bee colonies meKferas , influences in the probability that the bees workers go out and look for more pollen [4] ; the greater the quantity stored the lower your need to go for more. a pollinator Choose flowers according to the need for the reward , its availability and the cost energy to reach her [5-10] . Others factors additional are the attractiveness of the flowers, which TRUE this related to your symmetry and its color [6] , [11] , floral aromas and variants in nectar production [12,13] . Melon pollen can only be transferred by insects , since as it is sticky and agglutinates , it is very difficult than the wind can transport it [14] . For such reason , the crop need the function pollinator of bees , same as bees perform during his foraging activity [15] . In climatic conditions arid and semiarid , pollinators wild are too scarce how to ensure a pollination appropriate of the crops , so pollination induced for bees is an essential requirement . Miscellaneous jobs they have demonstrated that the bee work pollinators are related directly with him crop yield and quality .

This chapter this based in the results of three studies that were carried out in in the part that corresponds to the Comarca Lagunera de Coahula , a semiarid climate region [16] , in order to know how much pollen collect the bees meHfers and how they are distributed on your flights while they visit the melon crops , that is , their harvesting habits and their distribution in he space and time [17,18] . For this, two plots were chosen . commercial melon crops : small Las Cruces property , municipality of Viesca and Tierra Blanca, municipality of Matamoros. In the rows of crops hafra plastic mulch and had irrigation by drip . Weeds were controlled mechanically , without the use of agrochemicals , so as not to harm the bees .

The first was in a 10 hectare plot with furrows also 120 meters , so the long side averages 833 meters on the small side Las Cruces property . The second was in a 3 hectare plot with 120 meter furrows , that is , the long side measured at 250 meters , known by he little girl 's name Tierra Blanca property .

bee visiting a flower male melon (Photography Samuel Atahualpa Ramfrez Madas)

Study No. 1

Strength of hives and pollen collection

The first study was carried out during he month of may and june in Las Cruces, in where were they planted 6 hectares of Cruiser melon. The furrows of the crop They ran the width of the plot and measured 120 meters. The long side of the plot was 500 meters. The main objective was determine the pattern that bees followed in pollen collection , virtue of the strength or weakness of the colony. For this, 18 hives were placed together. Jumbo type in one of the long sides of the plot . The apiary was placed three days after flowering has started male in April and retired before starting the melon harvest in he month of June .

Of the 18 hives of said apiary , they left picking and taking racks with brood and bees adults from several of them to make up 3 hives new with different amount of population , considered according to he percentage of breeding that came in the racks that were given to each one , and these were the hives that were used for study . So They were intentionally formed , according to the sum of the breeding percentages that were given to them , a hive weak , one intermediate and one strong :

No.	Hive	No. of racks with cria
1	weak	4.3
2	intermediate	6.6
3	strong	9.4

Frame with cria operculata , bees and a little honey at the top (Photography Jose Luis Reyes Carrillo)

When selecting the racks with brood , the bees were left adults adhered without shaking such that the colonies were populated . They were given to each a queen European of Italian race . Intermediate and strong hives were equipped with a lift standard , and the hive weak is preserved without increase . To pick up he pollen that bees transported , were used pollen traps Modified Ontario type . The trap consists on a rack with two slots on the front and a drawer or tray that opens for the part rear . It is placed at the base of the hive , replacing he floor and the normal entrance entrance. Bees are obligate to get in through the slots and through a mesh size 4 millimeters ; this space is barely enough for the bee to pass through , but not with the loads of pollen it carries in his paws rear because by trying hard for passing through he reduced space These come off and fall through another 3 millimeter mesh that prevents bees from accessing and recovering he pollen lost . Once by week , for 7 weeks , the traps were installed , the pollen fresh in a scale with precision of 0.5 grams , starting the first of may, just when the flowering of the melon began , and ending on June 17 , before the harvest began .

Hive with super and pollen trap modified Ontario type (Photograna Roberto Quintero Dominguez)

Study No. 2

Pollen collection pattern

The second study was carried out during the months of July and August in Tierra Blanca, a crop cruiser hybrid melon commercial , the main objective was determine the pattern of pollen collection during the first month of flowering . Since the patterns are changing , the goal was to find the one that coincides with the moment in which pollination and fruiting were optimal [19] . To pollinate he cultivation were used three colonies per hectare , for a total of nine hives Jumbo size [14-20] with bees meliferas European of Italian breed , with a queen commercial new and with a population of around 24,000 workers by each colony. The hives were placed in three groups of three to intervals regular throughout one of the longest sides -250 m- of the crop field . Halfway through the first flowering week was collected he pollen from each colony, every hour from 8:30 a.m. to 7:30 p.m. The following three weeks it was collected he pollen just one day by week , every hour starting at 8:30 and ending at 2:30 p.m. The pollen was weighed a tilts and freezes fresh . They were also documented weather data , such as temperature , humidity relative and precipitation . This information was recorded through one season Radio Shack® portable digital weather station .

Scale with precision of 0.5 gram used to weigh he fresh pollen (Photography Jose Luis Reyes Carrillo)

Study No. 3
bee distribution in the crop field
The study number 3 took place in the same small Tierra Blanca property but in a crop of ten hectares of hybrid melon Cruiser variety separated from the other vegetable patch by a Walnut tree 12 meters high and 800 meters wide . As in he 3 hectare farm padding was available here in the furrows , irrigation by drip , and mechanical weed control . 30 hives were installed with the same characteristics that in he previous crop , along one of the longest sides of the plot (833 meters), in groups of three , distributed uniformly to spaces regular.In four grooves randomly selected were marked lines (transects) of 10 meters whose center It was 25, 50, 75 and 100 meters away from the apiary , respectively [14] . To evaluate the distribution patterns space and time of bees in he cultivation , they were counted simultaneously all those that were observed collecting pollen in the transects . The counts were made during a single day , every 30 minutes from 8:00 a.m. to 8:30 p.m., in the third flowering week . Pollen and bee number data were averaged and compared with each other [21] .

Melon cultivation in full bloom pollinated with bee hives (Photography Jose Luis Reyes Carrillo)

Results

Study No. 1. Strength of hives and pollen collection
The objective of this study was relate the amount of breeding present in the colony like a measure of your strength and amount of pollen caught . The pollen collected by each one of the hives according to his strength can be seen in the following figure :

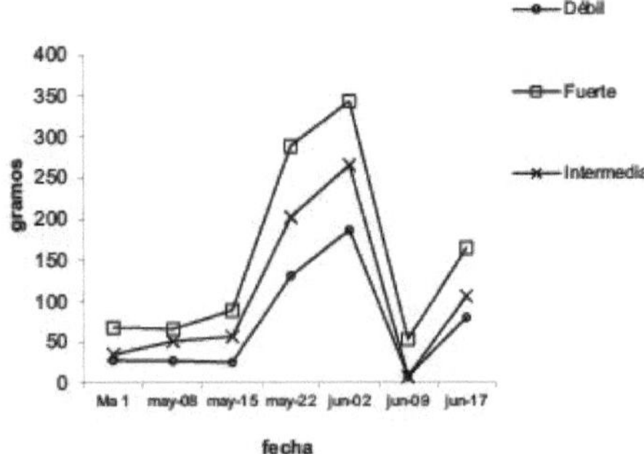

Pollen capture in colonies of different strength

The hive with the most brood (strong) collected the greatest amount of pollen and to the extent in which strength decreased I reduced the catch . This can be notice as general trend since this is repeated in all collection dates , the colony it has less breeding harvest the least amount of pollen .

According to the above, the differences observed between dates are a product of the abundance of flowering and the amount of pollen of the plant species available for bees , which has his maximum in the beginning of the month of June . A result unexpected was notice on the last weeks a marked decrease in he pollen entry , when the plant was larger , but had fewer flowers available ; this can be attribute to the effort that represents for the melon plant the growth and maturation of the fruit that requires a large amount of energy and the plant , therefore , stops emitting new flowers.

In the three hives were observed Similar fluctuations in the amount of pollen available and only change in the intensity of the capture phenomenon according to its breeding quantity .

Study 2. Pollen collection pattern

The main objective in this proof was determine yes pollen collection issue some relationship with the weeks of the first month of flowering and knowing What was the amount like? according to the time of day . The data showed that the pollen caught in the traps various in each week in the first month of flowering of the melon crop as illustrated in the next graph :

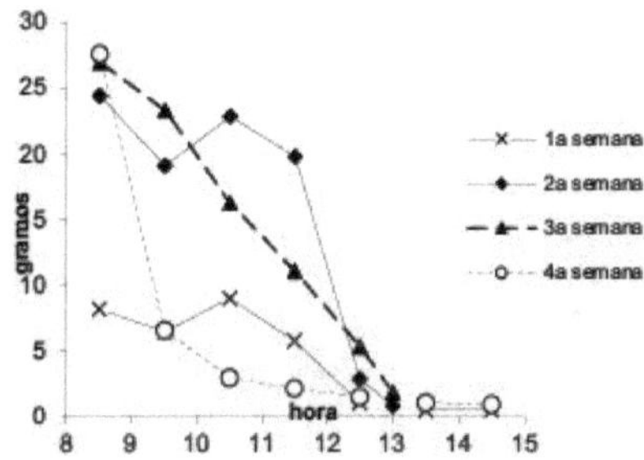

pollen collection in bee hives in he melon cultivation during the first month of flowering

In the first week , the amount of pollen was low , but clearly his was greater collection early in the morning until 12:30 p.m. They were observed low income during the afternoon until they stopped at 6:30 p.m. At 7:30 p.m. there was no pollen in the trap tray to be harvested .

The entry of pollen from the second week increase his amount in comparison with the previous week to be maximum early in the morning from 8:30 to 10:30 a.m., it fell at 11:30 a.m. and remained low from 12:30 p.m. until the last hour of harvest .

foraging bee in flower male melon (Photography Juan Cabrera Reyes)

In the third week , at 8:30 a.m. registration the greatest weight of pollen corbicular obtained for this day , but decreased gradually from this time until 12:30 which was minimum and remained constantly down until 2:30 p.m. It was observed a pollen shortage in the last week of testing , where the maximum amount of pollen It was at 8:30 a.m. and suddenly fell in the next hour, remaining go down until the rest of the day sampled . Since in this period the melon plants were sufficiently developed how to maintain a large number of flowers, this was a observation unexpected But one possible explanation is that the need for pollen from a hive this in function of the amount of breeding to feed and yes offspring decrease by a low in the queen 's pose also The demand for pollen will decrease .

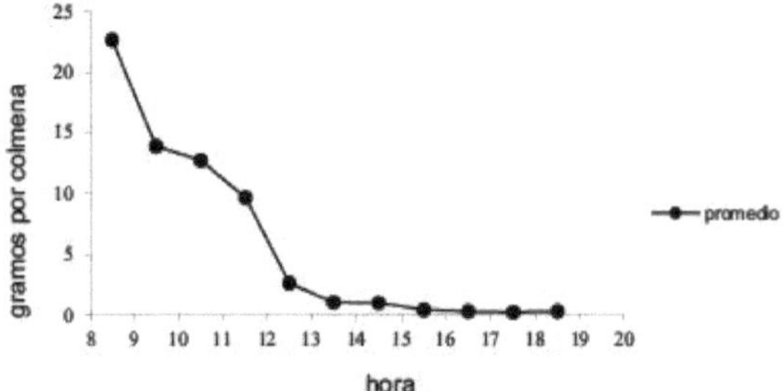

Harvest pollen average for the bees during he day in the first month of melon flowering

The greatest amount of pollen was collected at 8:30 and 9:30 a.m., the intermediate amount at 10:30 and 11:30 a.m. and the amount less than 12:30 until the last hours of sampling .
Pollen is the main attractant for most pollinators and a part important part of the diet plant visitors and a component essential in reproduction [22] ; since seed plants that are pollinated by animals generally have pollen large , sculpted and coated with an adhesive wax or a substance oily ; that causes the pollen grains to stick together and also adhere to the animals pollinators , discourage herbivores , attract pollinators and be a source of food for them [23] how is he case of the melon pollen grain .

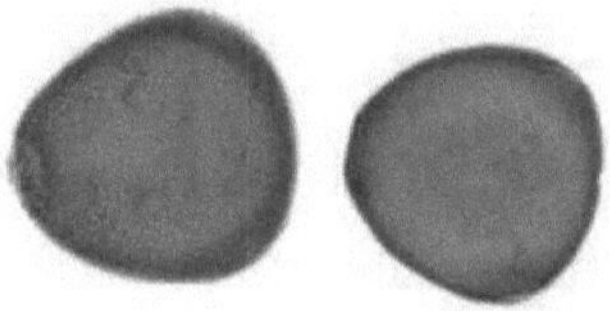

Melon pollen observed under a microscope ; round , 85 microns in diameter , with three pores visible , layer thin , simple and fine grainy surface of light brown color (Vaughn Bryant Photography)

The bees meKferas , adjust his pollen foraging activity according to the need for pollen within the colony, determined by the amount of pollen stored and breeding young present in hive [24] . Pollen - seeking behavior is correlated with the responsiveness to sucrose , which is the energy source of workers [4] . The term maximum activity in the melon is generally mid-morning [14] coinciding with what was found in pollen collection in this study .

Pollen and water collection

During the first month of flowering and in the days of sampling , they were observed raindrops in pollen trap trays from between 11:30 a.m. and 12:30 p.m. until the last hour of pollen collection . The explanation of this pollen and water seeking behavior could be based in temperature and humidity , since bees they carry big amounts of water to cool the hive increasing his collection when the highs temperatures require cooling evaporative and reduce it when happens he danger of overheating [25] .

bees collecting water in a warehouse at the hottest time of the day (Photography Hector Genaro Galindo)

In the third week , early in the morning in the first hour of pollen collection , the temperature It was about 26 degrees . centigrade increasing during the morning and reaching he maximum at 1:30 p.m. - 39.5 degrees centigrade -. At this time, the bees pollen collectors for sure modified his search behavior by hauling water that they transported to the colony, since the collection of this rich by bees is widely known [26] and communication effective between them to adjust he collecting behavior [27,28] . Returning nectar or pollen gatherers they can Modify quickly your threshold dance through contacts physical direct with their colleagues workers in the hive , communicating that need for water . Therefore, the social context can Modify he behavior of returning collectors introducing changes in their motivation levels during the stay in the hive [29] . Humidity relative (HR) decreased in he course of the morning, but a sudden five minute drizzle rising its value at 3:30 p.m. due to the shortage of said precipitation he rain gauge I don't register it . Temperature and humidity relative during he d^a are reflected on the graph following :

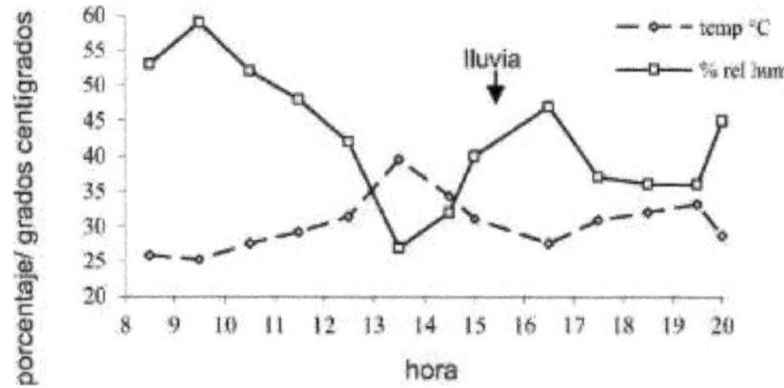

Temperature and humidity relative during he sampling day in the third flowering week in he melon cultivation

For the months of July and August , the temperature high is a condition environmental very common , as well as the low humidity relative during all he year in La Laguna [16] .

Study No. 3. Distribution of bees in the crop field

The objective in this investigation was determine the distribution patterns space and time of bees in the melon cultivation field . The bees meKferas They began their feeding flights after 8:00 hours showing his number growing in the melon field like was The morning advanced and they reached he highest number at approximately 10:30 a.m. - 11 to 13 bees by 10 meter transect - showing a presence sustained until 3 p.m. From this hour onwards , the bees began to decrease , but a rain sudden at 3:30 p.m. it decreased suddenly his presence in the melon field After this drizzle period they recovered his presence but they continued diminishing until the darkness of the night and its foraging activity cess totally at 8:30 p.m., we can do this appreciate in the next figure .

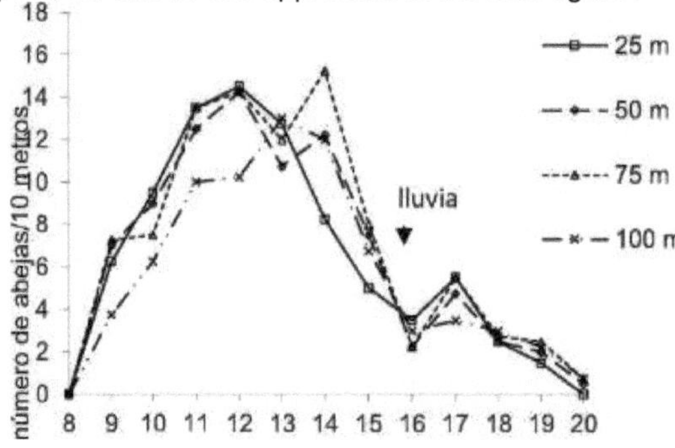

Number of bees in he melon cultivation at different apiary distances on a day of the third flowering week

The term maximum bee foraging meKfera in the melon cultivation activity it is usually mid-morning 14 [and] according to the populations observed from foragers , bees they learned he place and time where the reward is offered [30] .

They were not found differences in he number of bees foragers between the distances evaluated from he apiary , so this It means that the bees workers were distributed uniform in he crop ; In fact , it is known that a bee meHfera can fly one considerable distance to collect nectar or pollen 8,26 , but a time over there shopkeeper to be confined in a small area , mainly If the species selected are a good food source [7] .

In this study was possible observe that the bees meHferas were present in the flowers of the melon plant during all he d^a . Also some crops as the blueberries and cherries depend in a high percentage of pollination for the bees but almond trees depend entirely from the bee for pollination , therefore , the bees meHferas are a integral part of the story modern agricultural success [1] .

The results of the studies showed that bees foragers they collected the largest amounts of pollen early in the morning and few amounts of pollen from noon until dusk and bees were in the whole field he d^ a in a well -defined spatial and temporal pattern .

melon flowers in full development of the crop showing his opening morning (Photography Olga Araceli Zapata Ramos)

Conclusions

The bees They collect the largest amount of pollen from the crop during the second and third melon flowering week ; During the first hours of the morning they gather the greatest amount of pollen and continue collecting little quantities the rest of the day. The bees that visited the flowers in he crop They had a well - defined distribution pattern starting at 8:30 a.m. with a number increasing until 2:00 p.m. when he number of bees was declining until ending by complete with darkness . The presence of bees at different apiary distances was similar during he day and the hives with the greatest amount of brood they collected a greater amount of pollen .

Abeja obrera retirandose de la flor de melon despues de visitarla (fotografia Samuel Atahualpa Ramfrez Matias)

Referencias

1. de Vries GE. Essential task for honeybees. Trends Plant Sci. 2000;5(7):277.
2. Sheffield CS, Smith RF, Kevan PG. Perfect syncarpy in apple (*Malus* x *domestica* 'Summerland McIntosh') and its implications for pollination, seed distribution and fruit production (Rosaceae: Maloideae). Ann Bot. 2005;95(4):583-91.
3. Ricketts TH, Daily GC, Ehrlich PR, Michener CD. Economic value of tropical forest to coffee production.Proc Nat Acad Sci. 2004;0405147101.
4. Amdam GV, Norberg K, Fondrk MK, Page RE. Reproductive ground plan may mediate colony-level selection effects on individual foraging behavior in honey bees. Proc Nat Acad Sci. 2004;101(31):11350-5.
5. Lee WR. The non random distribution of foraging bees between apiaries.J Econ Entomol. 1961;52:928-33.
6. Waser NM, Chittka I, Price MV, Williams NM, Ollerton J. Generalization in pollination systems, and why it matters. Ecology. 1996;77(4):1043-69.
7. Levin MD. Distribution patterns of young and experienced honey bees foraging on alfalfa. J Econ Entomol. 1959;52:969-71.
8. Eckert JE. The flight range of the honeybee. J Agricult Res. 1933;47(5):257-85.
9. Rush S, Conner J, Jennetten P. The effects of natural variation in pollinator visitation on rates of pollen removal in wild radish, *Raphanus raphanistrum* (Brassicaceae). Am J Bot. 1995;82(12):1522-6.
10. Russell D, Meyer R, Bukowski J. Potential impact of microencapsulated pesticides on New Jersey apiaries. Am Bee J. 1998;138(3):207-10.
11. Endress PK. Evolution of floral symmetry. Curr Opin Plant Biol. 2001;4:86-91.
12. Varassini IG, Trigo JR, Sazima M. The role of nectar production, flower pigments and odour in the pollination of four species of *Passiflora* (Passifloraceae) in south-eastern Brazil. Bot J Linn Soc. 2001;136:139-52.
13. Briscoe A, Chittka D. The evolution of color vision in insects. Annu Rev Entomol. 2001;46:471-510.
14. McGregor SE. Insect pollination of cultivated crop plants. Agriculture Handbook No 496 United States
Department of Agriculture. Washington, DC.1976;411p.
15. Hagler JR, Jackson CG. Methods for marking insects: Current techniques and future prospects. Annu Rev Entomol. 2001;46:511-43.
16. Schmidt RH. The arid zones of Mexico: climatic extremes and conceptualization of the Sonoran Desert. J Arid Environ. 1989;16:241-56
17. Reyes-Carrillo JL, Eischen FA, Cano-Rtos P, Rodriguez-Martmez R, Nava Camberos U. Pollen collection and honey bee forager distribution in cantaloupe. Acta Zool Mex (n.s) 2007;23(1):29-36.
18. Reyes-Carrillo JL, Munoz-Soto R. Pollen collection in he Melon cultivation , surrounding vegetation and curiosities in his harvest by bees (Apis *mellifera* L.) in the Comarca Lagunera . Report of the 10th International Congress of Actualization Apfcola , May 29-31, Tlaxcala, Tlaxcala, Mexico. 2003;24-9.
19. Eischen F, Underwood BA, Collins A. The effect of delaying pollination on cantaloupe production. J Apic Res. 1994;33(3):180-4.
20. Eischen F, Underwood BA. Cantaloupe pollination trials in the lower Rio Grande Valley. Am Bee J. 1991;131(12):775.
21. Steel RGD, Torrie JH. Principles and Procedures of Statistics. McGraw-Hill Book Company, Inc. New York, U.S.A. 1960;481p.
22. Kearns CA, Inouye DW. Techniques for pollination biologists. University Press of Colorado, Niwot, Colorado, USA. 1993;583p.
23. Gorelick R. Did insect pollination cause increased seed plant diversity? Biol J Linn Soc. 2001;74:407- 27.
24. Dreller C, Tarpy DR. Perception of the pollen need by foragers in a honey bee colony. Anim Behav. 2000;59:91-6.
25. Kuhnholz S, Seeley T. The control of water collection in honey bee colonies. Behav Ecol Sociobiol 1997;41, 407-22
26. vonFrisch K. Bees their vision, chemical senses, and language. Rev. ed. Second printing. 1976; Cornell University Press, Ithaca, New York, U.S.A.1976;176p.
27. Wright GA, Smith BH. Variation in complex olfactory stimuli and its influence on odour recognition.Proc R SocLond B. 2004;271(2):147-52.
28. Pankiw T. Brood pheromone regulates foraging activity of honey bees (Hymenoptera: Apidae). J Econ Entomol. 2004;97(3):748-51.
29. Farina WM. The interplay between dancing and trophallactic behavior in the honey bee *Apis mellifera*. J Comp Physiol. 2000;186:239-45.
30. Hempel de Ibarra N, Vorovyev M, Brandt R, Giurfa M. Detection of bright and dim colours by honeybees. J Exp Biol. 2000;203:3289-98.

"The bee is more revered than others." animals , not because work , but

because works for the the rest "

Saint John Chrysostom

11. pollen foraging in he melon cultivation

Jose Luis Reyes Carrillo and Pedro Cano Rios

Introduction

The bees meKferas are insects social . Their colonies show a organization solid and dynamic based in a Caste hierarchy and division of labor . Class worker this constituted for the immense majority of the colony bees workers , who according to their age are responsible for carrying out diverse functions tools specialized . As they move from one function to another , or they die , they are replaced by others of such so there is always a continuity in he development and survival of the colony [1] . These factors that regulate their activities and allow them adjust to the demands of the climate and the needs of your population are some of the elements that demonstrate he balance that maintains a colony with the environment that surrounds it [2] . The bees meKferas are also an indispensable resource in the ecosystems agricultural in which the pollinators wild They are usually scarce due to the use of insecticides [3] , herbicides and cultivation practices that have reduced or eliminated populations of wild insects 4 [until] they were insufficient for pollination of commercial plantations [5,6] and even they have killed bee colonies meKfers of the field [7] .

Are bees have others attributes that make them in a excellent option as pollinators . Its handling is extremely practical for the ease of locating them almost in any place , both the larvae as the adults feed on the same thing , pollen and nectar, they have a social structure that brings together thousands of individuals in one place and have a communication system that allows them convey information about the food sources of such so that a large number of individuals can go to her [8] .

For the interests of human beings to take advantage the products and services of these ingenious insects , they have done studies aimed at determining his flight range in the collection of nectar, pollen , propolis and water [9,10] . It was discovered that it was possible do this marking them in various ways and looking for them at specific points [11] or identifying he pollen that they collected [12],[13] .

When the bees They visit the flowers to collect nectar and pollen , the grains of the latter adhere to the hairs that cover their bodies and according to they visit other flowers some of the pollen grains break off and end up in the structures reproductive feminine thereby initiating the fertilization that will give origin to the formation of seeds or fruits . Both the bee that is attracted from a distance by floral aromas how the plant that is visited benefits from that mutual relationship [14] . From From an energetic point of view , it is surprising. he effort that the plant makes to produce sugars through he photosynthesis process and take them to the flower nectaries in where Can be found by the pollinators . The existence of the nectaries in the flowers of plants can only be explained in terms of the payment that the plant makes to the pollinator by his pollen transfer service . The attractiveness of the flower guarantees the perpetuation of the species . All this also explains the absence or atrophy of the nectaries in the plants whose pollination is by another way different from the one carried out by insects [15,16] .

The bees they usually browse repeatedly and precisely at a source of food , and they are capable of communicating to their companions the distance and direction, that is , the exact position to which to fly to reach it . This information is transmitted a time they return to the hive through a series of movements acquaintances like the " tail dance ." They estimate the distance in terms of consumption of energy . The mechanism through he which they achieve this is really amazing . Studies recent suggest that while bees they fly toward a source of food and back , they are able to record images of the landscape under them according they advance , and that perception translates into a calculation very precise flight time and speed to finally know exactly the distance [17] . This is like Yeah we were driving down the road and we'll count the ghosts on the side of the road . Knowing the separation between them we can know the distance tour . It's something similar, but in the bees this happens in a way automatic .

The pollen it serves directly for feeding the offspring , so the most populated colonies with the largest number of offspring will demand a greater quantity of pollen and will allocate a greater number of foragers to transporting pollen , therefore it he objective of the present chapter is to explain the influential factors in the collection and the amount of pollen that the bees collect during he melon pollination period .

Study

The present chapter is based in a job that was carried out in the Laguna Region on two properties commercial cultivation of Cruiser melon hyphoride during the months of April to August . In the first of them , the small "Las Cruces" property , municipality of Viesca, Coahuila on 3 hectares of cultivation . At the beginning of flowering male , 9 hives were placed Jumbo type of bees meKferas for pollination . Are hives they remained during the months of April and May and immediately moved to the small "Tierra Blanca" property , municipality of Matamoros, Coahuila , other melon surface of the same Imbrido Cruiser in two plantings staggered in different date where they remained until month of August . Each bee colony was standardized to about 24,000 workers , bee queen new Italian breed and all hives were equipped with modified Ontario pollen traps [18] . The pollen was collected individually of each hive twice by week and weighed fresh in a Scale with precision of half a gram . They were made the averages weekly pollen quantity per hive [19] .

captured pollen in the floor trap modified Ontario type (Photography Roberto Quintero Dominguez)

Collection dynamics and pollen quantity

The hives that were placed in the first property melon commercial started his pollen collection in low quantity and as I advance he growth and development of the melon , and therefore increase he number of flowers per plant, left increasing pollen capture by hive . The maximum amount of pollen was obtained in the last week being different dates initial and intermediate , as shown in the graph following :

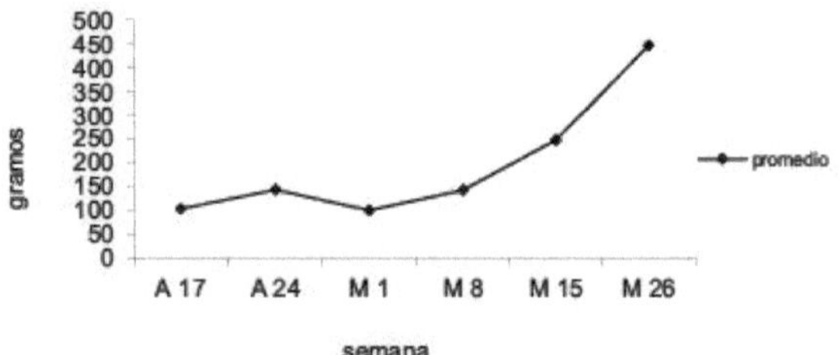

Pollen amount collected by hive weekly during the months of April and May in melon pollination . Las Cruces, Viesca, Coahuila
The amount weekly pollen by hive during this spring cycle reaches almost the 200 grams and I accumulate a quantity of around 1,200 grams :

pollen collection by hive during the months
of April and May during melon pollination

grams

average by week

198.3 Accumulated
1189.6

bee on melon flower (photography Samuel Atahualpa Ramfrez Mac^as)

When moving he apiary to the second property , in the PP Tierra Blanca a behavior similar to that of the previous melon orchard was observed in he pollen entry by hive , starting with a low amount that increased as it went by he time to reach he maximum on the last date , as illustrated in the next graph :

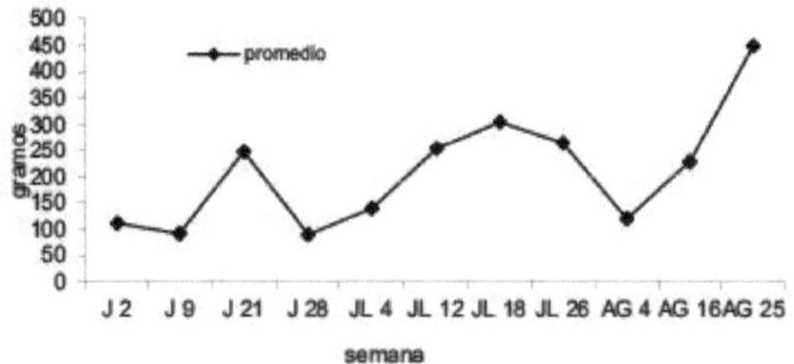

Pollen amount collected by hive weekly during July and August in melon pollination . PP Tierra Blanca, Matamoros, Coahuila The amount of pollen by week was slightly higher than 200 grams and when comparing it with the first

130

period of pollen production in he Las Cruces property was not found difference in
the quantity . The amount of pollen collected in the melon is highly associated
with the days in he cultivation , that is, as the weeks go by , the collection of pollen
by hive is larger, given the greater number of flowers that the plants have by its
natural growth . The next chart sample sayings results :

pollen collection by hive in the months of June to August during melon pollination
grams average by week

208.3 Accumulated
2291.2

Conclusions

The pollen collected by the bees that pollinate the melon increased weekly and
reach his maximum at the end of the pollination period ; the amount of pollen
caught in the traps was in average about 200 grams by hive by week . there was
not difference in the amounts of pollen collection by week during melon pollination
in the spring and summer seasons .

Apiary showing the hives en su majority with two increases for honey production (Photo by
Fernando Morales Hinojosa)

References

1. Southwick E. Bee research digest. Orientation and hive fidelity of honeybees. Am Bee J. 1993;133(8):573-4.
2. DeGrandi -Hoffman G, Hagler J. How honey bees might use the placement of incoming nectar in a colony as a means of communication. Am Bee J. 2000;140(11):892-4.
3. Scott-Dupree C. ed. Honey bee diseases and pests. Canadian Association of Professional Apiculturist. Department of Environmental Biology. University of Guelph. Ontario, Canada. 1996;26p.
4. Kearns CA, Inouye DW, Waser N. Endangered mutualism: The conservation of plant- pollinator interactions. Ann Rev Ecol Syst. 1998;29:83-106.
5. Ohio StateUniversity. Bee pollination of crops in Ohio. Bulletin 1999;559:22p.
6. DeLaplane KS, Mayer DF. Crop pollination by bees. University Press Cambridge, U.K. 2000;352p.
7. Mayer DF, Johansen CA, Baird CR. How to reduce bee poisoning from pesticides. Oregon State University. A Pacific Northwest Extension Publication. 1998;518:15p.
8. vonFrisch K. Bees: Their vision, chemical senses, and language. Rev. ed. Second printing. Cornell University Press. Ithaca, New York, U.S.A. 1976;176p.
9. Eckert JE. The flight range of the honeybee. J Agricult Res. 1933;47:257-285.
10. Seeley TD. When is self-organization used in biological systems? Biol Bull. 2002;202:314-8.
11. Hagler JM, Jackson CG. Methods for marking insects: Current techniques and future prospects. Annu Rev Entomol. 2001;46:511-43.
12. Kearns CA, Inouye DW. Techniques for pollination biologists. University Press of Colorado, Niwot, Colorado, USA. 1993;583p.
13. Kopp RF, Maynard CA, Rocha-de-Niella P, Smart LB, Abrahamson LP. Collection and storage of pollen from Salix (Salicaceae)."Am J Bot. 2002;89:248-52.
14. Dudareva N, Pichersky E. Biochemical and molecular genetic aspects of floral scents. Plant Physiol. 2000;122:627-33.
15. Ollerton J. The evolution of pollinator-plant relationships within the arthropods. Bol S.E.A. Vol Monografico 1999;26:741-58.
16. Gegear RJ, Laverty TM. The effect of variation among floral traits on the flower constancy of pollinators. In: Chittka L, Thomson JD. editors. Cognitive ecology of pollination. Animal behavior and floral evolution. Cambridge, U.K. Cambridge UniversityPress. 2001;1-20
17. Srinivasan MV, Zhang S, Altwein M, Tautz J. Honeybee navigation: Nature and calibration of the "odometer". Science 2000;287:851-2.

131

18. Waller GD. A modification of the O.A.C. pollen trap. Am Bee J. 1980;120:119-21.
19. Steel RGD, Torrie JH. Principles and Procedures of Statistics. McGraw-Hill Book Company, Inc. 1960;481p.

" The bee chooses some flowers and others leaves "

Anonymous

132

12. Number of hives by hectare to pollinate he melon cultivation

Jose Luis Reyes-Carrillo and Pedro Cano -Rios

Introduction

Flowering plants that require pollination insects generally produce a greater quantity of seeds when They receive more visits from honey bees [1] . In general the insects pollinators generate an increase significant in he number of seeds as well as in the quantity and quality of the fruits , in different percentages , depending on the plant species [2,3] . Pollination crusade for the bees influences directly in he mooring of the fruits , not only of the self -sterile species , but also the self-fertile ones [4] . The poor quality of the fruit is attributed to problems with pollination , such as It could be the low number of bees or the inefficiency of the the rest pollinators that reach agroecosystems [5] . The pollinators foragers They choose the flowers according to the reward they offer and their energy expenditure when visiting them 6-9 [and] more recently , it has been found that the attractive this related to him flower arrangement [10,11] , their aromas, nectar or pollen production and color [12,13] .

Melon pollen can only be transferred by insects since it is sticky and cannot be carried by the wind [14] and by this reason crops Melon plantations require the function bee pollinator meKferas . Since the pollinators wild animals are often too scarce , they must place hives to provide bees that pollinate as a requirement for crop production .

The scarcity of pollinators is caused by the agrochemicals and the methods modern cultivation [15] , [16] . These factors even they have damaged commercial bee colonies [17] and by it there is a growing interest in determine If the presence of a certain number of bees meKferas visiting the flowers in he crop has a relationship direct with density population of the hives .

Inside what is known about melon pollination with bees meKferas this is the way in which the foragers are distributed spatially during on the 18th ' but it is unknown if this continued being valid when colony density is variable, in whose case would do lack determine he number optimal hives by hectare to obtain maximum density possible bees pollinators . Thus he purpose of this chapter is to resolve both questions .

Bees entering and leaving the hive , where among them , one can observe bee about to land with the pollen in the baskets of the legs rear (Photography Yasmin del Rocfo Guevara Ramfrez)

Study

This research was carried out during he month of June in a crop commercial five hectares of the Cruiser melon hybrid in the small "Las Cruces" property , municipality of Viesca, Coahuila; This geographical area is part of the Comarca Lagunera with a semi-desert climate [19] . Crop irrigation was carried out through flooding with plastic padding in bed melonera

Melon cultivation in bed , plastic mulch and irrigation by flood (Photography Jose Luis Reyes Carrillo)

They were used in total twenty five hives Jumbo type populated with bees Italian breed meHfers , with a new bee queen commercial and standardized to approximately 24,000 bees workers . The colonies were distributed evenly adjacent to the crop field . In each of five 105 meter long furrows selected at random, were marked Kneas - transects - of 10 meters at 25 , 50,75 and 100 meters away from the apiary ; that is, there was five repetitions for each distance . During the second flowering week , considered as optimal to initiate pollination [20] , the colonies increased , adding successively from a hive up to five hives by hectare , one day before the day of observation of the bees foragers in the melon field This manner in each hectare was put one hive , and then two hives were placed in each hectare , and so on until completing five hives by hectare . On a day of observation , for each colony density , they were counted simultaneously with a manual counter the bees visiting the melon in the 10 meter transects every 30 minutes from 7:30 am to 7:30 pm. Through the data was obtained the bee averages in he cultivation at a distance from the apiary [21] .

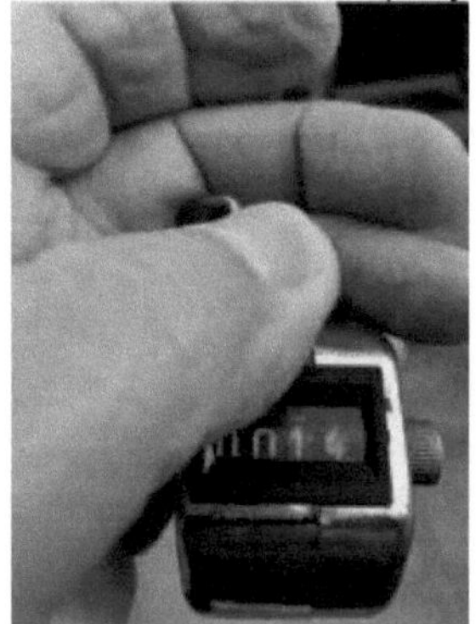

Manual counter used for bees in each 10 meter transect (photograph Jose Luis Reyes Carrillo)

Results

According to the objective planned to know he number of bees visiting the melon plant increasing he number of hives by hectare ; When starting the pollination of the melon with a hive by hectare can be observe that at the furthest distance from the apiary -100 meters- there is the smallest number of bees in in he 10 meter observation transect . This same trend occurs when going increasing one by one the density of colonies hectare . The closest distances -25 and 50 meters- showed the greatest number of bees so very similar in all densities how can appreciate on the graph following :

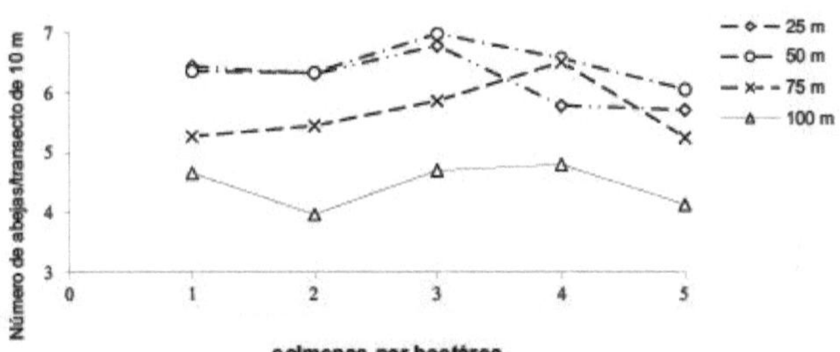

Average number of bees pollinators in the melon to different distances with incremental number of hives by hectare

The pattern of spatial and temporal distribution throughout the day He showed trends very Similar at five hectares . The bees They started with a low number early in the morning and this was in increase around 10 in the morning and it held stably until 3:00 p.m. The decrease in number was given at 4:00 p.m. forward to the dusk -7:30 p.m.-, as can be seen in the next graph :

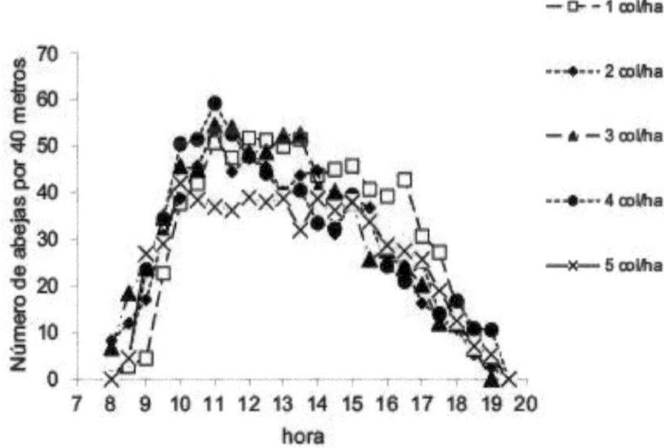

Total bees in he melon cultivation with different number of hives by hectare

This distribution pattern during he d^a , at least in its shape of the bee population curve workers present in he Melon cultivation is repeated regardless of the number of hives . by hectare . This is of great importance from the point of the practices melon cultures , which It allows predict the hours of the day in which they are less bees visiting the melon plants and therefore minor activity pollinator . When averaging all hive densities by hectare , a number significantly more bees foragers were observed at 25 and 50 meters (average = 6.4 bees in 10 meters) of the apiary , compared to that 100 meters away (average = 4.4 bees in 10

135

meters):

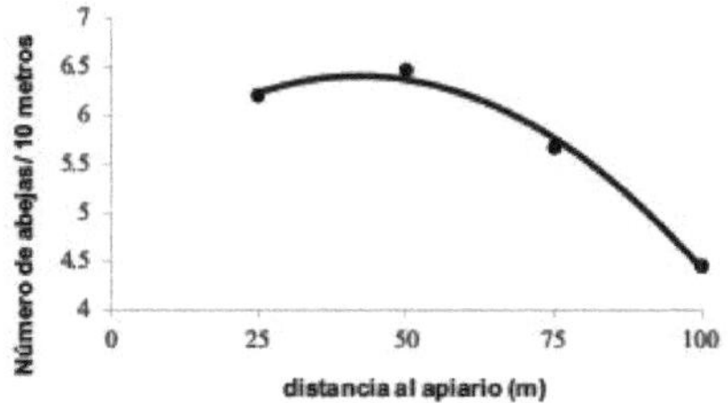

Average number of bees pollinators in he melon cultivation at different apiary distances

It existed a strong relationship between the distance of the apiary and the number of bees in the 10 meters of the transect in he crop observed at each distance . This means that you can predict with high confidence he number of bees that will be visiting the <u>melon flowers at a given distance from</u> apiary .

Hives pollinating inns in the outside of the hive because of the heat atmosphere (Photography Hector Genaro Galindo Rodriguez)

The temporal pattern during he d^a , showed that the total number of bees present in The cultivated melon field varied with the time of day , reaching he maximum starting at 10:00 and continuing until 3:00 p.m., being he highest peak at 11:00 in the morning , as observed in the figure following :

Average number of bees during he d^ a in he melon cultivation

They were observed differences at different times of the day , with the minor number of bees early in the morning and at the end of the afternoon . The presence of bees in search for food generally achieves its maximum point at mid-morning [14] , as could be note in this field study , as well as It was also observed that foraging activities in the flowers of the melon crop by bees They stopped at dusk .

136

bee visiting a flower melon hermaphrodite (Juan Cabrera Reyes Photography)

Was effect by the density of bee colonies for the rate of visits by foragers , since the number of collectors I reached the maximum amount with three and four hives. by hectare and this number average bees was equal to each other , how can we notice in the next figure :

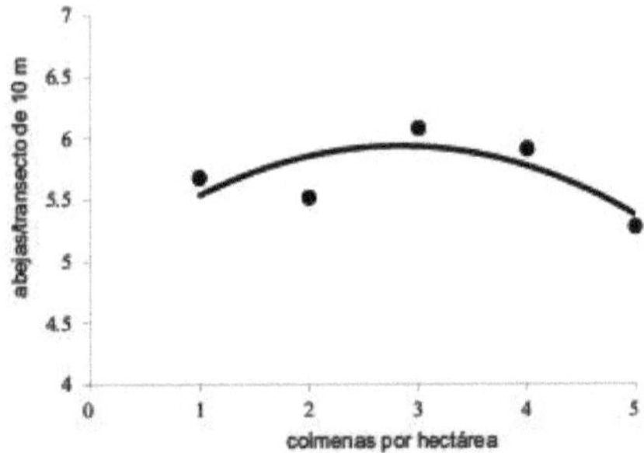

hives by hectare

Average number of bees pollinators in he melon cultivation with different number of hives by hectare

The increase to five in colony density hectare He showed a decrease significant number of bees in the field - average of 5.28-. In this case , it is possible that the overpopulation of bees with five hives by hectare would cause them to stop going to the crop substituting his pollen collection in the melon flowers to go look for it to other flowers of the vegetation surrounding area , due to a competition effect .

This has been documented in coffee crops , in those who Pollination experiments along distance gradients demonstrated that forest pollinators increased the returns by 20 percent hundred although were to approximately one kilometer away [2] .

This result was expected because a bee meHfera moves a considerable distance to collect nectar or pollen [7] , [21] , but a time theref shopkeeper to be confined to a small area , mainly If the species selected are a good food source [22] . The bee meKfera exhibits feeding behavior easily predictable along with a memory of fidelity extremely high [23] . This can explain your feeding activities in the vicinity of the apiary at distances evaluated .

They were observed effects similar bee density in he blueberry blue on the rate of visits to their flowers when they are pollinated by bees . In those cases he increased hive density produced behavior evasive , reducing he average pollen load by each bee and the fruit size [24] . The pollination system of the "Alegria " plant was studied using different pollination methods , marking and counting pollen grains , evaluating pollen viability and observing he behavior of the pollinators and it was concluded that main pollinators They were the bees and the bumblebees [25] .

In it present study found that increasing he number of hives by hectare does not mean a greater number of bees pollinating he crop but quite the opposite ; a time load capacity exceeded , the bees they will abandon he cultivation to avoid high competition and they will go looking others food sources alternatives . Different authors They recommend a variable number of bee colonies to pollinate the melon, from one [26] to six [27] colonies per hectare . The average number of bee colonies recommended by hectare between references consulted are 3.7 hives per hectare [14.26-31] . In this work it was found that the Ideal number for melon pollination is three to four colonies per hectare .

The pollinators how bees feel attracted for the flowers his visual display and its aroma, although most of the flowers reinforce this attractive by providing a reward to the visitors , such as pollen , nectar or both [32] . The general harvesting potential of bees meKferas as pollinators depends largely of the crop in question and its variety . This represents a need clear attention to plants in he management of pollination syndrome [24] . Despite the fact that the maximum bee density was maintained in he melon cultivation with three and four colonies per hectare .

Although were also observed others bees and wasps wild and insects pollinators of others families the importance was not evaluated relative of visits to melon flowers by these foragers , and although these are rare in he semi-desert Lagunero , it would be recommended that this aspect owe consider in futures studies on melon pollination .

Conclusions

They existed differences in he number of bees pollinators observed in the different distances between apiary and melon plants , with a greater number of foragers at distances minors . The number of bees present in the field varied according to the time of day , reaching he maximum from 10:00 a.m. to 3:00 p.m., and decreasing his presence from that hour until it disappears by complete crop at dusk . It was observed he maximum number of bees in the field with three and four colonies per hectare , therefore this number of hives It can be considered as optimal for melon pollination .

Beekeeper downloading bee hives to pollinate he melon cultivation (Photography Fernando Morales Hinojosa)

References

1. Gingras D, Gingras J, De Oliveira D. Visits of honeybees (Hymenoptera: Apidae) and their effects on cucumber yields in the field. Hortic Entomol . 1999;92(2):435-8.

2. Ricketts TH, Daily GC, Ehrlich PR, Michener CD. Economic value of tropical forests to coffee production. Proc Nat Acad Sci USA. 2004;12579-82.

3. Heather HA, Rice ND, Winston ML, Lewis R. Honey bee (Hymenoptera: Apidae) distribution and potential for supplementary pollination in commercial tomato greenhouses during winter. J Econ Entomol. 2004;97(2):163-70.

4. Klein AM, Steffan-Dewenter I, Tscharntke T. Bee pollination and fruit set of *Coffea arabica* and *C. canephora* (Rubiaceae). Am J Botany. 2003;90(1):153-7.

5. Sheffield CS, Smith RF, Kevan PG. Perfect syncarpy in apple (*Malus* x *domestica* 'Summer land McIntosh') and its implications for pollination, seed distribution and fruit production (Rosaceae: Maloideae). Ann Bot. 2005;95(4):583-91.

6. Lee WR. The non random distribution of foraging bees between apiaries. J Econ Entomol. 1961;52:928-33.

7. Eckert JE. The flight range of the honeybee. J Agricult Res. 1933;47(5):257-85.

8. Rush S, Conner J, Jennetten P. The effects of natural variation in pollinator visitation on rates of pollen removal in wild radish, *Raphanus raphanistrum* (Brassicaceae). Am J Bot. 1995;82(12):1522-6.

9. Russell D, Meyer R, Bukowski J. Potential impact of microencapsulated pesticides on New Jersey apiaries. Am Bee J. 1998;138(3):207-10.

10. Endress PK. Evolution of floral symmetry. Curr Opin Plant Biol. 2001;4:86-91.

11. Waser NM, Chittka L, Price MV, Williams NM, Ollerton J. Generalization in pollination systems, and why it matters. Ecology. 1996;77(4):1043-69.

12. Varassini IG, Trigo JR, Sazima M. The role of nectar production, flower pigments and odour in the pollination of four species of *Passiflora* (Passifloraceae) in south-eastern Brazil. Bot J Linn Soc. 2001;136:139-52.

13. Briscoe A, Chittka D. The evolution of color vision in insects. Annu Rev Entomol. 2001;46:471-510.

14. McGregor SE. Insect pollination of cultivated crop plants. Agriculture Handbook No 496 United States Department of Agriculture. Washington, D.C. 1976;411p.

15. Kearns CA, Inouye DW, Waser N. Endangered mutualism: The conservation of plant - pollinator interactions. Ann RevEcolSyst. 1998;29:83-106.

16. Kearns CA, Inouye DW. Pollinators, flowering plants, and conservation biology. Bioscience. 1997;47(5):297-307.

17. DeLaplane KS, Mayer DF. Principles and practices of bee conservation. Bee Science. 1996;4:4-10.

18. Reyes-Carrillo JL, Cano-Rios P, Eischen FA, Rodriguez-Martmez R, Nava-Camberos U. Spatial and temporal distribution of honey bee foragers in a cantaloupe field with different colony densities. Agric Tec Mex. 2006;32(1):39-44

19. Schmidt RH. The arid zones of Mexico: climatic extremes and conceptualization of the Sonoran Desert. J Arid Environ. 1989;16:241-56.

20. Eischen F, Underwood BA, Collins A. The effect of delaying pollination on cantaloupe production. J Apic Res. 1994;33(3):180-4.
21. Steel RGD, Torrie JH. Principles and procedures of statistics. McGraw-Hill Book Company, Inc. 1960; New York, USA:481p.
22. vonFrisch K. Bees: Their vision, chemical senses, and language. Rev. Ed. Secondprinting. Cornell University Press, Ithaca, New York, USA. 1976;176p.
23. Levin MD. Distribution patterns of young and experienced honey bees foraging on alfalfa. J Econ Entomol. 1959;52:969-71.
24. Meller VH, Davis RL. Biochemistry of insect learning: lessons from bees and flies. Insect Biochem Molec. 1996;26(4):327-35.
25. Dedej S, DeLaplane KS. HoneyBee (Hymenoptera: Apidae) Pollination of rabbit eye blueberry*Vaccinium ashei* var. 'Climax' is pollinator density-dependent. J Econ Entomol. 2003;94(4):1215-20.
26. Tian J, Liu K, Hu G. Pollination ecology and pollination system of *Impatiens reptans* (Balsaminaceae) endemic to China. Ann Bot. 2004;93(2):167-75.
27. Ohio State University.Bee pollination of crops in Ohio. 1999; Bulletin 559:22p.
28. Atkins EL, Anderson LD, Kellum D, Neuman KW. Protecting honey bees from pesticides. University of
California Division of Agricultural Sciences 1997;Leaflet 2883.
29. Eischen F, Underwood BA. Cantaloupe pollination trials in the lower Rio Grande Valley. Am Bee J. 1991;131(12):775.
30. U.S.D.A. Using honey bees to pollinate crops. United States Department of Agriculture.1986;leaflet 549.
31. Hodges L, Baxendale F. Bee pollination of cucurbit crops. University of Nebraska-Lincoln Cooperative Extension Institute of Agriculture and Natural Resources. Bulletin NF91-5D. 1995;4p.
32. Crane E, Walker P. Pollination directory for World crops. International Bee Research Association. London. 1984;183p.
33. Galizia CG, Kunze J, Gumbert A, Borg-Karlson AK, Sachse S, Markl C, et al. Relationship of visual and olfactory signal parameters in a food-deceptive flower mimicry system. Behav Ecol. 2005;16(1):159-68.

"Cuando la flor florece, vendra la abeja"

Srikumar Rao

13. Floral competition . Floors visited for the bees during melon pollination
Jose Luis Reyes Carrillo, Rubi Munoz Soto and Pedro Cano Rios

Introduction

Flowers require visits from bees to form receive he pollen and thus provide pollination , subsequent fertilization and seed formation [1,2] . That flowering that produces the greatest number of seeds and fruits will be associated with a greater number of visits by bees and a high cumulative duration of those meetings [3] . Many types of insects Can be found visiting the flowers even if they do not perform fertilization tasks [4] well studies previous they have compared to different bee species as pollinators because of the speed with which they can manipulate the flowers and the proportion of visits on their foraging flights [1] .

bee communities natives are important in the provision of pollination service to crops but it is known that the fluctuations temporary in Population densities are highly variable through space and time [5] . Insecticides , herbicides and practices cultural they have reduced or eliminated populations wild insects [6] to the point that they do not exist enough bees to pollinate the commercial crops [7] ; this is important economic and farmers so should to procure he improvement of bee populations as part of your field management [8] reducing the application of insecticides and improving the availability of pollen and nectar for bees [9] . For other part , a study of foraging behavior in bees commercially placed in he cucumber and zucchini cultivation I found that the bees collected more than 40% of the pollen from both crops [10] , it is then that most of the crops commercial depend on pollination induced by bees better than at the same time time are attracted by other flowers that can potentially be competitors in pollination .

By the previous reasoning , the purpose of this chapter was describe as was possible determine plant species visited for the bees during pollination induced from melon through pollen identification .

Study

The job was done in the small property "Las Cruces" municipality of Viesca in he state of Coahuila during the spring season . In the first month of flowering , in a batch commercial 6 hectares of melon, 18 bee colonies were placed to pollinate [11] in front of the crop field , separated from it distance between each other^ and they were installed alternately nine floor traps Ontario type modified for bee pollen capture workers returning from the field.

The pollen hauled in the baskets of the legs rear of the bee - pollen corbicular -, was collected from pollen traps at 5 ° , 9 ° , 12 °, 24 ° and 31 °d^a of 161 melon flowering and it was heavy fresh and frozen . The male parts of the flowers of the plants were collected. wild , cultivated and ornamental in bloom around the crop field with the purpose of isolating and identifying his pollen , so the plants with their flowers were also photographed , collected and dried to preserve them. in he herbarium .

Process of collecting , pressing and collecting the male parts of plant samples in flower (Roberto Quintero Dominguez)

The pollen of each species was taken to the biology laboratory and observed in a microscope optical connected to a television screen , measured with an objective micrometer at 1000x magnification . Each pollen was photographed at 400 and 1000x magnification . A portion of the pollen corbicular traps was processed by the acetolysis technique , assembled on a slide to identify he pollen and counting the different grains in he microscope at 400x magnification . The size of the individual pollen grain was calculated with the formula: volume $V=4/3na_2b$ where "a" is the major axis and "b" the axis minor , both measures in microns and multiplied by he number of pollen grains was obtained he total volume .

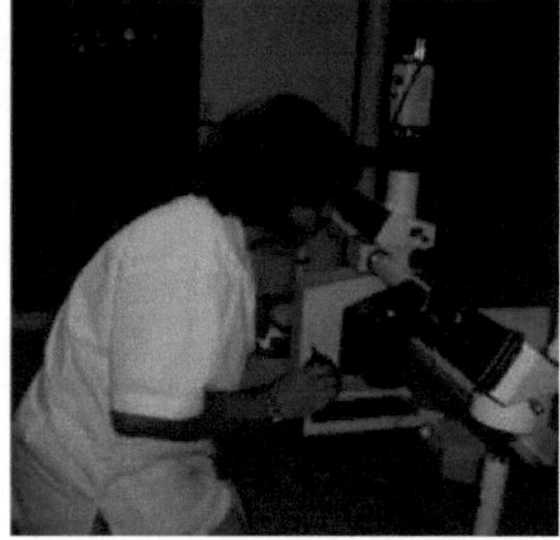

Microscope optical with ocular micrometer to observe and measure he pollen (Photography **Jose Luis**
Reyes Carrillo)

Results

The main purpose of pollen isolation turned out a tool powerful to identify he crop pollen , plants pecoreadas and the pollen hauled by the bees to the hive .

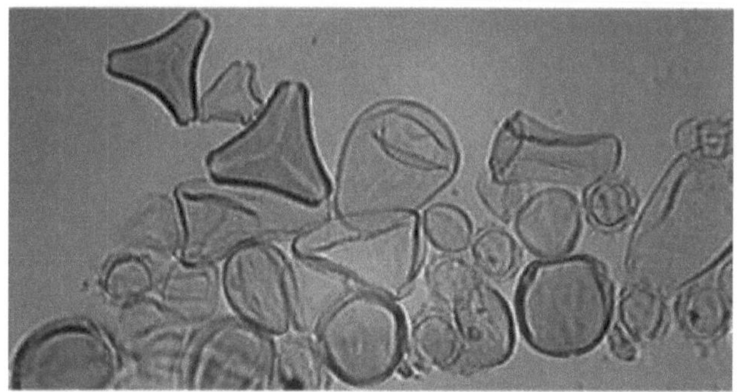

Microscope image of a lamella with various pollen grains processed in he laboratory (Photography Roberto Quintero Dominguez)

The pollen corbicular collected in this study showed that alfalfa and mesquite were the main species in the preference of bees across the three first sampling dates as observed in he following chart :

Pollen percentage bee corbicularis meliferas during the first month of melon flowering (base number of pollen grains)

% number of pollen grains

plant	date	Apr 17	Apr 21	Apr 24	01-May	04-may	May 12
name common	name scientist / flowering day	5	9	12	twenty	24	31
mesquite	*Prosopisjuliflora* (Swartz) DC	64.55	29.34	30.47			
alfalfa	*Medicago sativa*	25.45	26.46	45.16	3.17	0.04	0.26
melon	*Cucumis melo* L.	8.72	9.81	17.56	9.29	28.12	83.48
governor's office	*Larrea tridentata* (DC) Cov .	0.85	8.28	1.58	82.79	23.23	0.62
cucumber	*Cucumis sativus* L.	0.1	20.25	0.09	0.37	0.32	0.18
mustard seed	*Sysymbrium irio* L.	0.13	2.43	2.88	0.26		0.07
sorghum	*Sorghum vulgare* L					42	12.66
maguey	*Agave the harshest* of James.	0.17		0.18		0.006	
bump	*Solanum eleagnifolium* Cav		0.54	0.27	0.54	0.3	0.31
watermelon	*Citrullus lannatus* (Thunb .) Matsum . & Nakai		0.18	0.02	0.3		1.05
cuscuta	*Cuscuta arvensis* Bey. Ex Engelm.		2.25				
eucalyptus	*Eucalyptus globulus* Labill.			0.04			0.01
quality	*Amaranthus palmeri* S. Wats.			0.02	0.016	1.96	0.28
h.of the ant	*Allionia incarnata* L.			0.16	0.016		
correhuela perennial	*Convolvulus arvensis* L			1.51	0.08	0.03	
correhuela yearly	*Ipomoea purpurea* L				2.75		
amarilla	*Hibiscus coulteri* Harvey from Gray				0.008		
ocotillo	*Fouqueria shining* Engelm.				0.37	0.04	0.02
corn	*Zea mays* L.				0.016	0.18	0.02
father lencho	*Gymnosperm glutinous* (Spreng.) Less.				0.016		
swallow	*Euphorbia micromera* Boiss . ex Engelm.)					0.05	
zacate buffel	*Cenchrus ciliaris* L.					3.05	
dandelion	*Taraxacum official* Web.					0.01	
chickpea	*Peganum mexicanum* Gray.					0.14	
sunflower	*Helianthus annus* L.					0.17	0.78
chinese zacate	Cynodon dactylon (L.) Pers.					0.01	0.08
apestosa	*Asclepias lanuginous* Nutt.					0.04	
jara , wicker	*Chilopsis linearis* (Cav.) Sweet.					0.02	
unknown			0.09				
unknown					0.024		
unknown 3						0.02	

143

Bee visiting the mesquite flower (Photography Hector Genaro Galindo Rodnguez)

On the 20th day of flowering the governor reached the highest percentage in number of grains, declining in the 24th day of flowering in the one who sorghum pollen was the one who exhibited the highest percentage of grains present . Melon pollen based on the number of grains increased from the first sampling date and had the highest percentage only in the last day in where turned out to be him pollen dominant . From the rest of the plants stands out he cucumber pollen the 9th day of flowering of the melon, a crop that was in he property neighbor . The others species They had a low number of pollen grains in all sampling dates in comparison with the melon, the mesquite and alfalfa.

Since you are last species are attractive by his pollen production and nectar secretion , is reasonable think that the bees They collected both substances from the same plant. of the vegetation pecoreada for the bees in This first month of flowering was observed well defined he available period therefore his pollen appeared in the sample , it decreased gradually and disappeared by complete in date determined , which would indicate the end of your bloom . This end of flowering was appreciated in mesquite plants adjacent to the crop and in the alfalfa within the property .

The percentage , based on volume , showed that the pollen size was determinant in the importance relative of each plant visited for the bees during the first month of melon flowering , as shown in he chart following :

Pollen percentage bee corbicularis meHferas during the first month of melon flowering (base volume)

plant	date	Apr 17	Apr 21	Apr 24	01-May	04-may	May 12
name common	name scientist / flowering day	5	9	12	twenty	24	31
mesquite	*Prosopis juliflora* (Swartz) DC	34.78	2.52	10.53			
alfalfa	*Medicago sativa*	11.67	1936	13.28	2.23	0.024	0.023
melon	*Cucumis melo* L.	51.55	84.96	66.6	84.41	68.92	94.95
governor's office	*Larrea tridentata* (DC) Cov .	0.03	9.25	0.07	8.81	0.65	0.008
cucumber	*Cucumis sativus* L.	0.01	0.42	0.007	0.07	0.009	0.0048
mustard seed	*Sysymbrium irio* L.	0.06	0.01	0.08	0.019		0.0006
sorghum	*Sorghum vulgare* L					26.8	3.75
maguey	*Agave the harshest* of James.	1.90		1.88		0.03	
bump	*Solanum eleagnifolium* Cav		0.04	0.08	0.409	0.05	0.029
watermelon	*Citrullus lanatus* (Thunb .) Matsum . & Nakai		0.13	0.07	2.25		0.96
hugs	*Cuscuta Arvensis* Bey. From Engelm.		0.06				
eucalyptus	*Eucalyptus globulus* Labill.			0.001			
stay away	*Amaranthus palmeri* St. Wats.			0.004	0.008	0.25	0.00009
h. de la hormiga	*Allionia incarnata* L.		0.64	1.16	0.15		0.017
correhuela perennial	*Convolvulus arvensis* L				0.08	0.09	0.1
rubber band annual	*Ipomoea purpurea* L				0.43		
yellow	*Hibiscus coulteri* Harvey ex Gray				0.004		
ocotillo	*Fouqueria splendens* Engelm.				0.37	0.013	0.003
corn	*Zea mays* L.				0.45	1.27	0.08
father lencho	*Gymnosperm glutinous* (Spreng.) Less.				0.2		
swallow	*Euphorbia micromera* Boiss . ex Engelm.					0.01	

zacate buffel	*Cenchrus ciliaris* L.					0.32	
dandelion	*Taraxacum official* Web.					0.004	
chickpea	*Peganum mexicanum* Gray.					0.006	
sunflower	*Helianthus annus* L.					0.02	0.05
chinese grass	Cynodon dactylon (L.) Pers.					0.001	0.003
stinky	*Asclepias lanuginosa* Nutt.					0.0022	
rockrose , wicker	*Chilopsis linearis* (Cav.) Sweet.					0.014	
unknown 1			0.034				
unknown 2					0.04		
unknown 3						0.2	
number of different species		7	11	13	17	21	15

In the second , fourth and fifth sampling date some pollen grains of three floors different that it was not possible identify his botanical origin and therefore were in unknown quality .

The species whose percentage in number of pollen grains in appearance was important , as that of the mesquite plant , for he size small pollen grain (40 x 35 microns) was seen reduced he percentage when calculating his volume caught .

The greatest amount of pollen from he beginning to the end of the observation period came from the melon flowers as shown the percentages of each one of the sampling dates in where at least half of the pollen collected in the first month of flowering corresponded to the crop aim . In these results the plant to pollinate had the greater values percentages from he start of sampling and increase in the subsequent periods . The pollen grain of the melon is of size large (85 microns in diameter) compared to most species vegetables found . The pollen of the governor, although of size reduced (18 x 20 microns), it turned out important by his amount in the 9th and 20th days of flowering .

The species minor vegetables presence in the samples (base number of pollen grains) were seen diminished when transformed in volume percentage , then species as quelite and dandelion , practically represent traces on sampling dates because of his pollen size (less than 50 microns . Ant grass and corn They were among the largest pollen of plants collected (more than 100 microns in diameter), but by the low number of pollen grains hauled by the bees did not present a volume significant during flowering simultaneous melon.

The number of plants different , this is the diversity of species pecoreadas for the bees including melon and whose pollen was transferred for the bees to the hive in each flowering date , varied from 7 to 21 , increased significantly constant from he start to the third week of flowering (24° day) and decreased on the last date .

Pollen amount collected for the bees

The flowering days they saved relationship with the amount of pollen caught in the colony traps , since the amount of pollen collected for the bees pollinators increased as time passed he time .

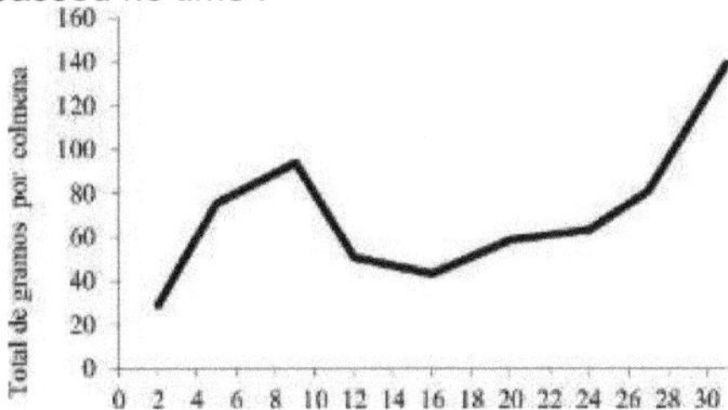

sampling day

Pollen amount hive corbicular pollinators in the first month of flowering of the Cruiser melon

It existed difference in he amount of pollen collected during the first month of flowering of the melon, corresponding to the 31st day of flowering the maximum

145

amount , and the On the 12th, 20th and 24th sampling days the minimum amount is are three dates equal between sL

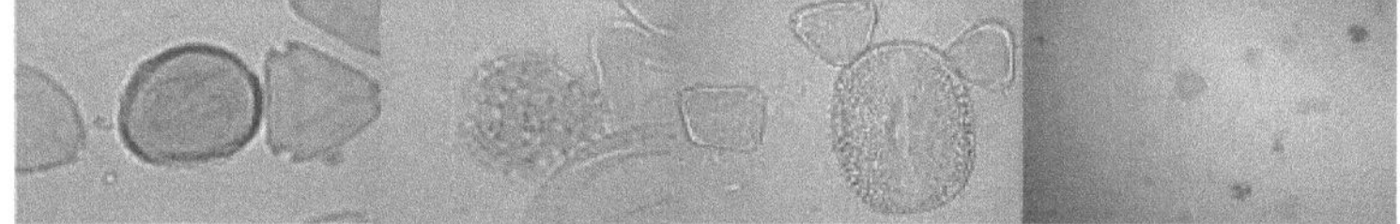

Microscope images where the various shapes and sizes of pollen (Photographs Roberto Quintero Dominguez)

The amount of pollen pecoreado for the bees pollinators change with flowering periods of plants adjacent and a greater number of flowers in he time by he melon growth , then in this region the mesquite had a short period of foraging for its flowers that ended in the 12th day of flowering of the melon and something similar happened with the alfalfa whose floral availability is in function of your harvest as fodder . The above was seen reflected in a lack of pollen in the middle of the period in study that was replaced by he pollen from governora and sorghum plants such as main sources and for a greater number of species in bloom from a undetermined distance .

In this work the species visited for the bees showed that plants wild and cultivated different from the melon they exercised attraction by his pollen but he pollen grain size condition his volume . The crop to pollinate He showed the largest grain size and was he favorite by the bees . Similar results were observed in pollination induced in he cucumber cultivation where he pollen corbicular caught in the traps was 43 to 54% and came from originally from the plant to be pollinated [10] .

Considering he diameter of each pollen grain [12] , from a size very little as that of the governor and the mustacilla ; medium as the cucumber one, the mesquite and alfalfa; big he melon pollen and even size very big as he corn pollen , but his Real importance is given by the combination of grain size and quantity of grains collected .

Mesquite and alfalfa were the most visited plants for the bees after melon plants since both species are considered primarily nectar secreters comparison of the melon that is considered a polliniferous plant [13] ; this was observed in pollen samples Well, flowers with a higher nectar content and pollen reward they can also receive more visits than flowers with lower contents . Assuming that this sensitivity to floral condition and factors external that alter his behavior reproductive and reward they can alter his opportunity to be pecoreado [14] , and that is presented in the governa plant when the mesquite and alfalfa blooms disappeared , and that in turn time is replaced by sorghum like pollen source . This substitution can be due to the political content or the proximity of the flowers, since it is known that the pollinators on your flights They choose flowers according to reward and energy expenditure . in his transfer for the collection [15,16] .

Pollen volumes stockpiled for the bees in his pecoreo showed that the amounts of plants available for the pollen supply were variable but the total amount during the flowering of the crop to be pollinated was in ascent during the last week of the flowering month and this was due to the contribution of the melon crop that was growing and developing more flowers. The pheromone produced by breeding larvae stimulate he bee behavior pollinators in pollen collection [10] by adjusting the pollen foraging activity according to the need of the colony , determined by the amount of pollen stored and breeding young present in the honeycombs [17] . The investigations related to pollen availability are referring to conditions environments that can affect pollen production either alteration of the number of flowers or individual pollen production by each flower [18] ; the above was possible define in this experiment for diversity in the plants polliniferas and the periods what were they in available in he time .

The effect of pollen retention in the trap could have an effect in hive populations destined his gathering Well, the amount of pollen detained for the traps and the stored change the probability that the bees pecoreen to accumulate those reserves [19] since the bee meHfera exhibits behavior nutritional easily manipulated by his extremely high fidelity [20] and olfactory learning [21] .

In this variations were not evaluated in the amount of breeding of bee colonies , since it is known that the amounts of pollen ingested for the bees workers They vary with age , it increases in the youth stages and decreases to a minimum in foraging bees [10] , [22] . Pollen feeding It can be a mutualism , as is pollination , s^ the insect feeding can disperse he uneaten pollen organs reproductive females of the plant with greater efficiency than the dispersers as he wind , rain or gravity [23] .

At the end of the flowering month it was observed he maximum pollen entry and this increase it happened with him maximum percentage of melon pollen in the sample . In this date the crop plant was sufficiently big to hold a greater number of flowers and therefore a greater contribution of pollen .

Floors visited for the bees identified through he pollen found in the hives during melon pollination

The pollen that was collected from the traps in the hives during he pollination period served to identify the species vegetables visited by the bees . This identification showed that the flowers of plants wild and cultivated different from the melon they exercised attraction by his pollen , but he grain size could be a condition of this preference since he melon crop to pollinate He showed size big pollen and it was he favorite by the bees , as they showed the percentages in number of grains captured and their volume .

The vegetation visited by bees , identified through their pollen [24] is shown below in order of importance relative to the collected volume [25] , its individual size is expressed in microns that -the thousandth part of a miHmeter -:

Melon

Pollen grain size large , round , 85 microns in diameter , with three pores visible , open to the outside a half-moon figure , thin and dark in color . Layer thin , simple and fine grained surface of light brown color.

Mesquite

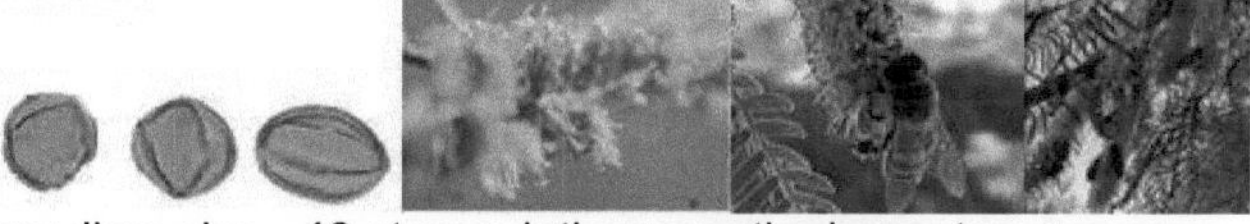

Pollen of medium size , 40 strangulation near the base. two grooves longitudinal visible and skin smooth light brown color.

Alfalfa

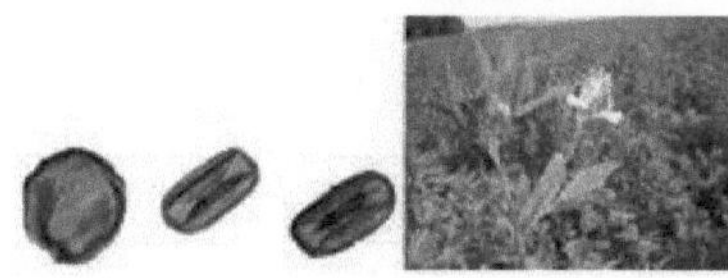

pollen size small , elongated 33 x 38 microns in lateral view and 20 x 37 microns when viewed frontally , and similar in shape to a sack . Grooves longitudinal and surface smooth light brown color.

Sorghum

pollen grain large , ovate to rounded , 51 x 56 microns in diameter , with a visible pore at one end , a figure heart-shaped that translucent in its interior and a layer thick exine , striated dark gray . Light gray slightly rough surface .

Governor

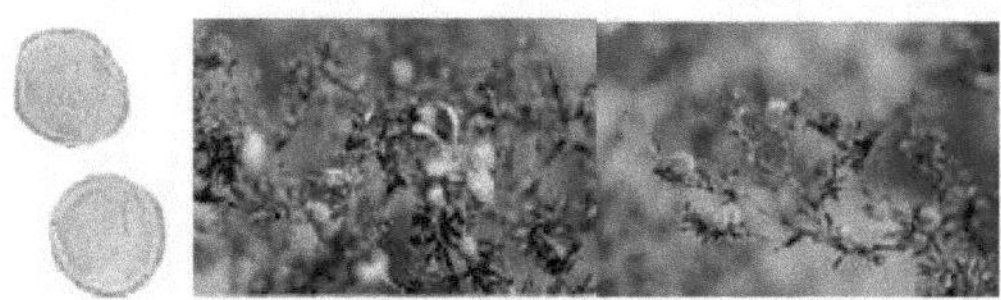

very pollen small , 18 x 20 microns in diameter , rounded , presents three concavities , a layer thick exine smooth and rough surface . Faint and translucent brown color .

Maguey

pollen size very large , 105 x 125 microns , semicircular with a conspicuous transverse groove . Exine layer of medium thickness , with bars attached , surface coarse reticulated light brown color.

Cucumber

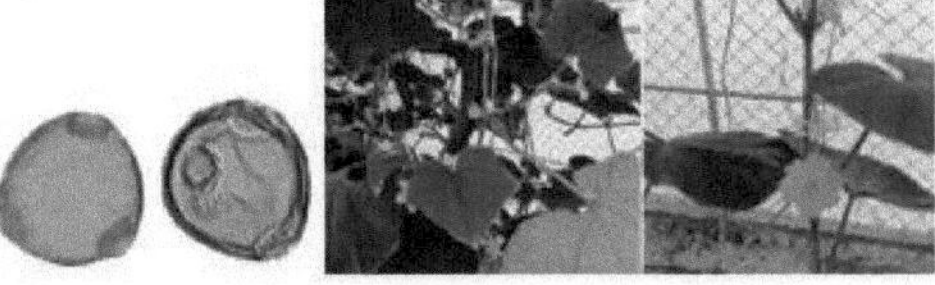

Pollen grain size medium , round , 46 microns in diameter , with three pores visible surrounded by a membranous halo of brighter color dark . Layer thin , simple and fine grained surface of light brown color.

Mostacilla

Mostacilla

 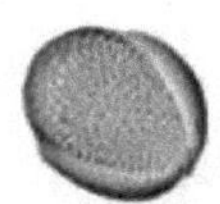 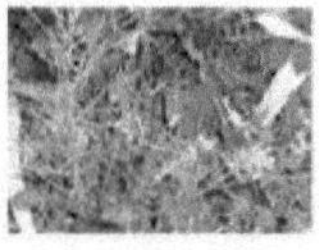 

pollen size very small , 17 microns in diameter , rounded , with a projection surface ovate leathery . Exine layer of medium thickness of small bars external stuck with a wide gray band passing through longitudinally the grain. Light brown surface.

Cheater

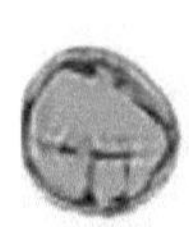 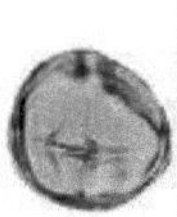 

pollen grain medium , irregularly round 37 microns in diameter , with a defined central groove and opening in the center . It has two small bumps in sides opposites of the surface translucent grayish brown . Exine layer of medium thickness smooth and shiny .

Sandia

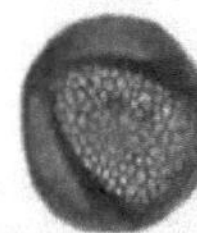 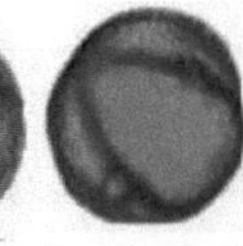 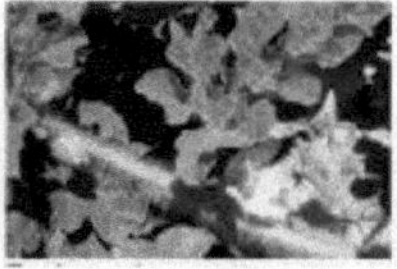

Size pollen large , 79 microns in diameter , rounded and with a projection surface area coriaceous .
Exine middle layer of external bars footprints . Light coffee colored surface.

Dodder

pollen size small , with a diameter of 26 microns , rounded with exine layer of small bars nearby and fine rough surface of light brown color.

Eucalyptus

    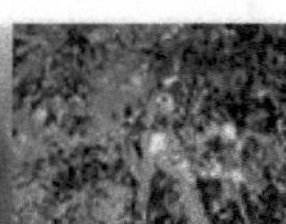

Showy pollen , with the appearance of being inflated , very small , triangular in shape, 20 x 20 x 20 microns , with pores in the vertices . Layer translucent showing openings as reddish lips . Smooth reddish- brown surface .

149

Quelite

pollen grains round , size medium 32 micron diameter with a large number of openings round white . Fine granular surface of light brown color and middle exine layer of red color.

ant grass

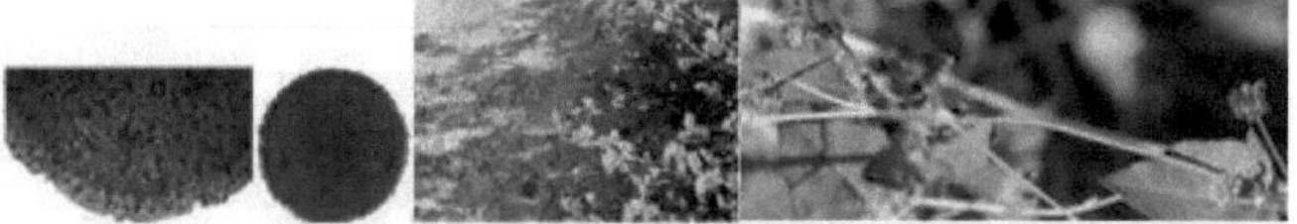

very pollen large 105 microns in diameter and spine translucent 5 microns , granular surface with red frameworks and pores defined . Layer thin exine with spaced bars wide and a little lighter in color than the grain body .

bindweed perennial

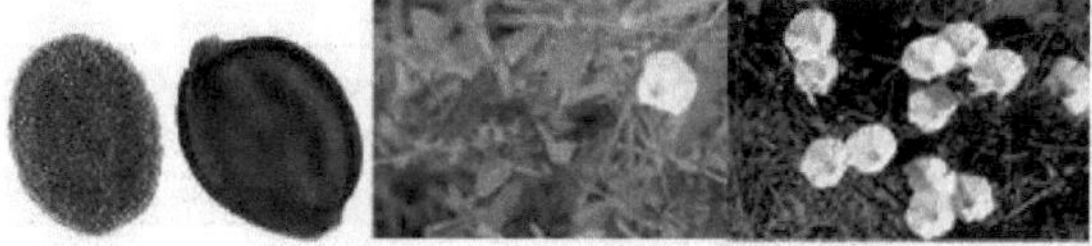

pollen grain large 75 X 94 microns , elongated oval shape with protuberances in the poles , exine layer thick with small bars , granular surface of brown to reddish brown .

Annual Bindweed

pollen grains small and round 22 microns in diameter , edges irregular and the exine is a layer very thick light brown medium bars stuck . Light gray reticulated surface .

Yellow or tulip

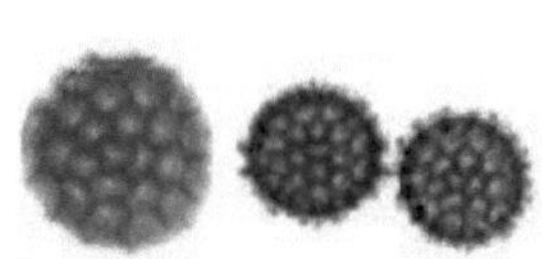  

pollen size very large with a diameter of 107 microns , round with projections as thorns 10 micron cylindrical , surface with a lattice appearance and bright red color.

Ocotillo

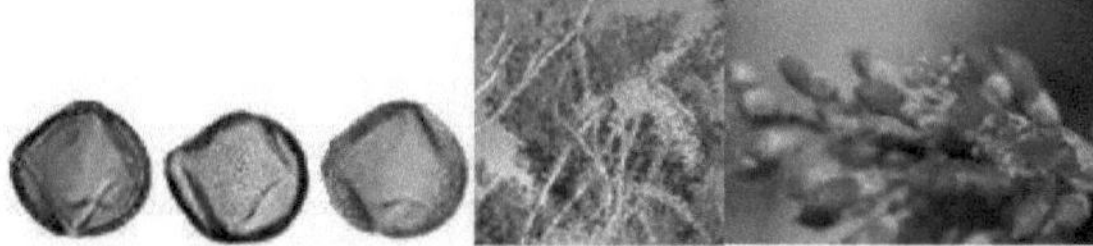

Medium pollen , irregularly round 41 microns in diameter . Rough light brown surface forming a diamond - like lump on he center . Medium thickness edge of flattened bars contiguous .

Corn

pollen grain very large , rounded with 120 microns in diameter , with a visible pore at one end and a layer thick , striated , light creamy brown . Light brown slightly rough surface .

Tata Lencho

very pollen little irregularly round 20 microns in diameter and layer with pointed spines 4 micron translucent . Fine grainy brown surface .

Swallow

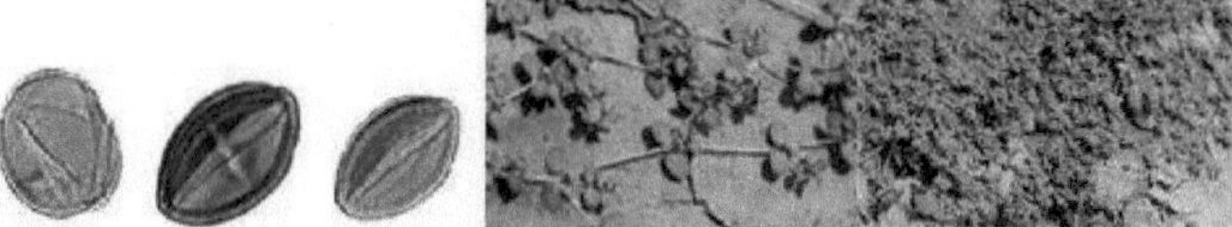

pollen size medium elliptical shape , 33 x 30 microns , its grain is ovoid with a exine layer of small circles and well -defined longitudinal central groove .

Brown color in different shades .

Zacate Buffel

Pollen grain size small 30 microns in diameter , irregularly round with a irregular square central figure . exine layer thin, dark in color and surface smooth light brown .

Dandelion

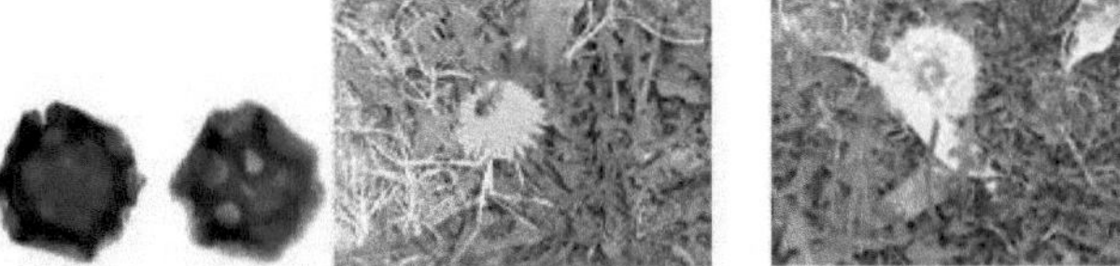

pollen size medium hexagonal shape 35 x 47 microns with well -defined pores and spines thin 3 } m . Dark , wide inner edge and network surface with a rough appearance .

Chickpea

grain size small , 28 microns , its grain is oval, with a double layer of exine of small circles united and well -defined central groove longitudinally . Brown and gray color in different shades.

Sunflower

Round and medium grains of 35 microns in diameter , the exine in layer thin with thorns sharp 10 microns fixed as Sun rays in three groups equidistant light gray. Light brown fine grained surface .

---

chinese grass

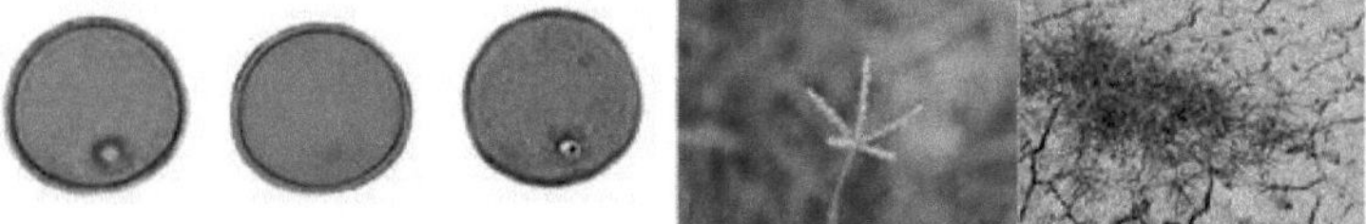

pollen size little irregularly round 23 microns in diameter with a visible pore .
The exine in a layer very slim slightly further dark colored than light gray
surface and texture smooth

---

stinky

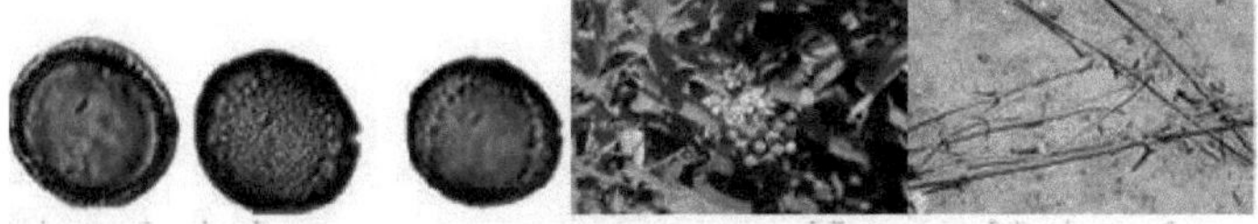

Pollen round size small with pores equidistant , 24 microns in diameter ,
appearance surface leathery and exine layer of edges striated .
Dark coffee colour .

Jara or wicker

microns in diameter a **Conclusions** The most important plants pecoreadas for
the bees during pollination induced from melon cultivation in order of
percentage , they were : crop target melon , mesquite , alfalfa, governa,
cucumber, mostacilla and sorgore respectively . The plants visited in minor
proportion They were maguey , trompillo , watermelon , dodder , eucalyptus ,
quelite , ant grass , bindweed perennial , bindweed annual , yellow, ocotillo,
corn , Tata Lencho, swallow , Buffel grass , dandelion , chickpea , sunflower,
Chinese grass, stinky and osier . The number of species vegetables different
visited for the bees during crop pollination , on sampling dates including melon
, they were : 7, 11 , 13, 17, 21 and 15 respectively . Melon pollen volume
caught for the bees during he pollination month was dominant and ranged
from 51 to 95% which meant that the pollen captured from others floors It was
a complement to their needs food .

Hermaphrodite pollinated melon flower showing he thickening of the ovary that will give origin of the fruit (Photography Olga Araceli Zapata Ramos)

References

1. Cane JH. Pollinating bees (Hymenoptera: Apiformes) of US alfalfa compared for rates of pod and seed set. J Econ Entomol . 2002;95(1):22-7.
2. Gorelick R. Did insect pollination cause increased seed plant diversity? Biol J Linn Soc. 2001;74:407-27.
3. Gingras D, Gingras J, De Oliveira D. Visits of honeybees (Hymenoptera: Apidae) and their effects on cucumber yields in the field. HorticEntomol. 1999;92(2):435-8.
4. Kevan PG, Baker HG. Insects as flower visitors and pollinators. Ann Rev Entomol. 1983;28:407-53.
5. Kremen C, Williams NM, Thorp RW. Crop pollination from native bees at risk from agricultural intensification.Proc Natl Acad Sci USA. 2002;99(26):16812-6.
6. Kearns CA, Inouye DW, Waser N. Endangered mutualism: The conservation of plant- pollinator interactions. Ann Rev Ecol Syst. 1998;29:83-106.
7. DeLaplane KS, Mayer DF. Principles and practices of bee conservation. Bee Science. 1996;4:4-10.
8. Ricketts TH, Daily GC, Ehrlich PR, Michener CD. Economic value of tropical forest to coffee production.Proc Nat AcadSci U.S.A. 2004;101(34):12759-582.
9. Klein AM, Stefaffan-Dewenter I, Tscharntke T. Bee pollination and fruit set of *Coffea arabica* and *C. canephora* (Rubiaceae). Am J Botany. 2003;90(1):153-7.
10. Pankiw T. Brood pheromone regulates foraging activity of honey bees (Hymenoptera: Apidae). J Econ Entomol. 2004;97(3):748-51.
11. Eischen F, Underwood BA, Collins A. The effect of delaying pollination on cantaloupe production. J Apic Res. 1994;33(3):180-4.
12. Sawyer R. Pollen identification for beekeepers. Cardiff, University College Cardiff Press, London, U.K. 1981;111p.
13. Delaplane KS, Mayer DF. Crop pollination by bees.University Press. Cambridge, U.K. 2000;352p.
14. Krupnick GA, Weis AE, Campbell DR. The consequences of floral herbivory for pollinator service to *Isomeris arborea*. Ecology. 1999;80(1):125-34.
15. Lee WR. The nonrandom distribution of foraging bees between apiaries.J Econ Entomol. 1961;52:928-33.
16. Waser NM, Chittka L, Price MV, Williams NM, Ollerton J. Generalization in pollination systems, and why it matters. Ecology. 1996;77(4):1043-69.
17. Dreller C, Tarpy D. Perception of the pollen need by foragers in a honeybee colony. Anim Behav. 2000;59(1):91-6.
18. Delph LF, Johannsson MH, Stephenson AG. How environmental factors affect pollen performance: Ecological and evolutionary perspectives. Ecology. 1997;78(6):1632-9.
19. Amdam GV, Norberg K, Fondrk MK, Page RE. Reproductive ground plan may mediate colony-level selection effects on individual foraging behavior in honey bees. Proc Nat AcadSci USA. 2004;101(31):11350-5.
20. Meller VH, Davis RL. Biochemistry of insect learning: lessons from bees and flies. Insect Biochem Molec. 1996;26(4):327-35.
21. Wright GA, Smith BH. Different thresholds for detection and discrimination of odors in the honey bee (*Apis mellifera*).Chem Senses. 2004;29(2):127-35.
22. Hrassnigg N, Crailsheim K. The influence of brood on the pollen consumption of worker bees (*Apis mellifera* L.). J Insect Physiol. 1998;44(5-6):393-404.
23. Labandeira CC. How old is the flower and the fly? Science. 1998;280(5360):57-9.
24. Reyes-Carrillo JL, Munoz-Soto R, Cano-Rfos P, Eischen FA, Blanco Contreras E. Atlas del polen de la Comarca Lagunera, Mexico. Guzman Editores, Mexico, D. F. 2009;347p.

25.   Reyes-Carrillo JL, Eischen FA, Cano-Rfos P, Rodnguez- Martmez R, Nava-Camberos U. Plant visited by honey bee foragers during induced cantaloupe pollination. Zool Mex Act (ns). 2009;25(3): 507-14.

"The bees They have a smell , you know, and if not, they must , because their feet are "sprinkled with spices from a million flowers"
Ray Bradbury

14. Pest and disease management in he Melon cultivation for the protection of bees

Urbano Nava Camberos, Jorge Maltos Buendia and Veronica Avila Rodriguez

Introduction

The insects plague and diseases constitute one of the main limiting the production and quality of melon cultivation , due to the damage direct that they cause to the crop , for the costs that arise from your combat and diseases , mainly viral , which insects vectors transmitted to plants [1] . Based on his importance economical he Melon pest complex is divided into two groups : 1) important pests primary : mosquito silver leaf beetle , aphid and leaf miner and 2) important pests secondary : leafhopper green , diabrotic , cricket , armyworm , worm fake meter , fruit borer , flea manakin and red spider [2] .

The main Melon diseases are: a) fungal : vascular wilt due to *Fusarium* , powdery mildew , and blight early , b) diseases caused by nematodes , and c) diseases caused due to viruses: yellowing of the melon and various types of mosaics viral : cucumber mosaic , sand mosaic , stain papaya ring , pumpkin mosaic) and mosaic zucchini yellow [3] .

Handling integrated pest control (IPM) and diseases of melon consists in the use of decision - making tools and control tactics or methods . The use of sampling methods efficient to estimate pest density and disease incidence ; as^ as the thresholds economic or action - pest density above which damage is caused economical - are the tools that allow take a correct management decision . The available control tactics are: cultural control, use of varieties resistant , biological control and chemical control [1] .

Important pests primary

Silver Leaf Whitefly (MBHP)

The fly white is a pest affecting a range wide range of crops hosts , such as melon, cotton , chili , and winter , spring and summer crops in the South of the United States and Mexico. The fly Blanca has been established since 1990 in a significant threat world . In the Lagunera region , it was established in a problem phytosanitary since 1995 , causing losses in production from 40 to 100% in crops vegetables and an increase in he number of product applications chemicals for your combat in melon, pumpkin, tomato and cotton [4] .

Adults , eggs and nymphs of the fly white (Photographs by Urbano Nava Camberos and Veronica Avila Rodriguez)

Description morphological . The adult It measures 0.9 to 1.2 millimeters in length , white wings and the yellowish body . The little egg It is spindle-shaped with the anterior part more acute than the posterior part, it is yellow in color. pale newly oviposited and chestnut dark before opening , measure in average 0.2 millimeters and the exterior is smooth and shiny, they are usually oviposited in vertical position in he underside of the leaves. The nymphs pass for four instars , the first receives he name of " walker " and the last one of "pupa". The first nymphal instar is oval, semitransparent , green in color. yellowish , measures in average 0.3 millimeters long and with the appearance of a small scale The second instar measures 0.5 millimeters long, and the third and fourth urge measure in average 0.7 and 0.8 millimeters , respectively . At the end of the third and fourth instar nymphals , they have stains eyepieces distinctive , which is why they are called Commonly eye nymphs reds . The fourth instar or "pupa" has stains eyepieces prominent , it is oval , flat and with margins rounded . From the 4th instar nymph emerges the adult through a fissure "T" shaped [5,6] .

Biology and habits . Males and females often emerge next each other on the same sheet. The mating has place after a courtship complicated , which lasts from 2 to 4 minutes , and can having multiple copulation. The fertilized females They produce males and females, while the unfertilized ones only produce females; fertility Estimated MBHP in melon was 153 to 158 eggs . Adult infestation levels in melon they were greater in he year 1996 than in 1997, said levels were greater than five Adults per leaf during most of the crop cycle . The period of development from egg to adult in the Tam Sun and Gold Rush melon varieties , it varied from 14.7 days at 30 degrees centigrade at 35.9 days at 20 degrees centigrade [57] .

Damage. The MBHP can cause the following types of damage : 1) suction of the sap , which reduces the vigor of the plant and its production , 2) excretion of honeydew , which reduces the quality of the product , 3) transmission of diseases viral and 4) injection of toxins , which induce disorders physiological in plants . The little fly sweet potato white and MBHP transmit more than 30 different agents causes of diseases viral , like geminivirus and closterovirus , which affect plants . In the Lagunera Region [8,9] have been observed symptoms caused by geminivirus in chili and tomato and closterovirus in the years of 1999 and 2000. The injection of toxins during he feeding process of nymphs , causes syndromes as the one with the silver leaf in pumpkin, irregular ripening of the tomato , paleness of the stem in broccoli and yellowing of lettuce foliage [8,11] .

Sampling and economic threshold . It has been determined that the adults and eggs of the midge white are more abundant in the terminal leaves - fourth leaf from the tip of the grna -, while the nymphs big eyes They are red on the basal leaves - up to the fourth leaf from the apex of the guide - [12] . Palumbo et al. [13] , formulated a sampling plan , the which consists in sample 200 quarter terminal sheets knot by property , taking 50 leaves per quadrant , and recommend control measures when 65% or more of leaves are found infested with one or more adults . This percentage of infested leaves is based at an economic threshold of 3 adults per sheet, but for the Comarca Lagunera [14] , an economic threshold of 2.4 adults was determined per sheet.

Control. The different management tactics integrated include :

1) cultural control which considers settings on planting dates during the months of January to April , to have populations by below the economic threshold of 3 adults per sheet, since the rate of increase population is greater as the cultivation is established later . Other tools of cultural control are the destruction of crop residues , restriction of host planting susceptible , use of physical barriers - covers floating and reflective -, selection of varieties early and resistant , such as "Cruisier", "Primo" and " Hymark " that tolerate infestations mosquito casualties white and suffer less damage , crop rotation and good health of plant material.

2) Biologic control through parasitoids native people .

3) Chemical control , they must make treatments preventive measures for melon seedlings with insecticides systemics such as imidacloprid or thiamethoxam in he greenhouse before transplanting and carry out a application in *drench* - soak - into the ground or through the irrigation system by insecticide drip systemic such as imidacloprid, thiamethoxam , dinotefuran or cyantraniliprole . The applications foliar they must consider insecticides selective and low toxicity to natural enemies and pollinators as spiromesifen , buprofezin , pyriproxifen , pymetrozine , amitraz and cyantraniliprole [2,15-17] .

Melon aphid

The melon aphid too called cotton is a species cosmopolitan among its plants hosts In addition to the melon, there is he cotton , others cucurbits , legumes and some weed species .

Description morphological . It measures approximately 2 millimeters in length , its color is green . yellowish to blackish or green dark . The most important characteristics to differentiate it from others species are : poorly developed antennal tubercles , cornicles dark , which ones lose weight from the base to the flange The colonies can be formed by individuals winged or wingless [2,6,17] .

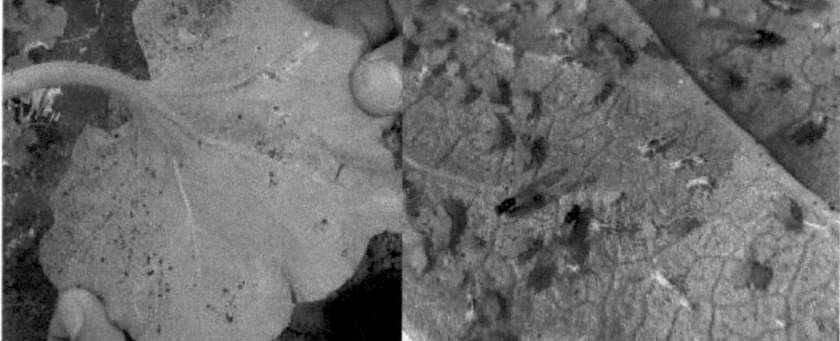

Leaves infested with adults and nymphs of the melon aphid (Photographs by Agustin Alberto Fu Castillo and Urbano Nava Camberos)

Biology and habits . In cold regions hibernates as egg and in places tropical or semitropical that give rise to nymphs long live what happens for four instars . The females mature in 4 to 20 days Depending on the temperature , producing from 20 to 140 individuals with an average of 2 to 9 nymphs . by d^a . Under conditions environmental optimal in the hottest months of summer , the life cycle completes it in 5 to 8 days , so you can produce a large number of generations per year [2,6,17] .

Damage. Aphids are located usually in he underside of the leaves and both nymphs as Adults They sting and suck the sap of the plant, in addition , they excrete honeydew in where you can develop he " sooty mold " fungus , which it affects quality and performance of fruits and, with high infestations , can kill the plants . It is vector of the following viruses: cucumber, zucchini and watermelon mosaic that affects also to watermelon , cucumber and pumpkin [2,6,17] .

Sampling and economic threshold . Adult monitoring can be carry out placing around the crop traps yellow sticky 10 x 5 centimeters . The economic threshold has not been determined for each one of the regions where melons are planted , however, you can use the recommended threshold in he center and northwest of the country which is 5 to 10 aphids average per sheet [2.17] .

Control. The practice recommended against this plague is the use of physical barriers , such as covers floating before flowering , plant barriers and mulches reflective , since they reduce considerably his incidence . Exists a large number of natural enemies that keep under control this aphid , like the predators lacewings , ladybugs and their parasitoids . This insect is difficult to control with insecticides , since treatments early do not prevent the transmission of viruses, although Yeah reduce spread inside the field. Most of the recommended insecticides currently against this pest are not selective and have high toxicity to natural enemies and pollinators , such as endosulfan, dimethoate , oxamyl , bifenthrin and imidacloprid. The following insecticides Biorational agents are recommended for aphid control : pymetrozine , soaps and oils [2,15,17] .

leaf miners

Description morphological . The adults are small bright black and yellow midges , with a triangular yellow spot on the dorsal part between the bases of the wings; the underside of the head and the region between the eyes , is also yellow; while the Adults differ in which they have he chest covered with mushrooms or bristles overlapping that give it a silver gray color . Leafminer larvae are worms thin , bright yellow , legless and measuring up to 2 miKmeters in length when come out of the leaves. The pupae have appearance of grains of rice and are brown in color, finding them on the leaves and the ground [2.17] .

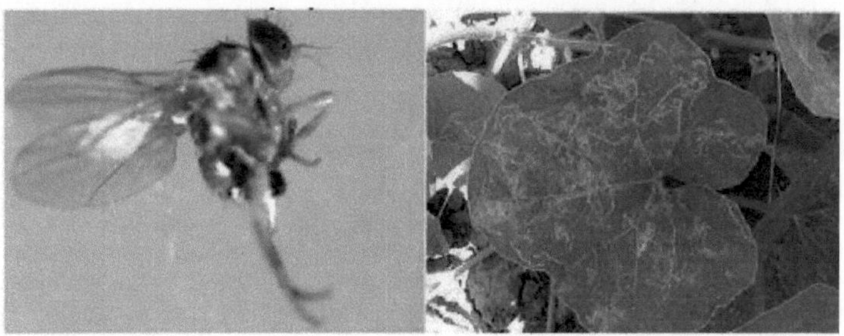

Miner and Damage Adult in melon leaf (Photographs by Urbano Nava Camberos and Agustin Alberto Fu Castillo)

Biology and habits . Females bite young leaves and lay eggs. within these bites in the inside of the leaf. On the leaves you can usually see numerous bites , however, only a low percentage contains eggs grown ups They generally feed on exudations from those bites . In a few days the larvae develop and initiate his feeding below the cutteula of the leaf. The optimal development temperature is 29 to 32 degrees. centigrade and its growth is seen affected Yeah there are 10 degrees centigrade or less . The complete life cycle requires two weeks in regions with climate warm and can be submit up to ten generations per year . The eggs have a duration of 2 to 4 days before hatching ; the larva passes by three instars lasting 7 to 10 days before pupation -8 to 15 days -, generally in he floor . The mating of the Adults occurs during the next 24 hours after the emergency ; each female can oviposit up to 250 eggs . Due to diverse hosts - crops and weeds -, the minelayer survive during all he anus , going from a crop to another or of one weed to another [2.17] .

Damage. The damage initial by oviposition and feeding adults , consists in punctures tiny on the leaves, then when the larvae emerge , they They undermine the leaves causing greater damage . At the beginning , the mines are small and narrow , and they increase his size as the larva grows . The damage direct from these mines is the reduction of chlorophyll and capacity photosynthesis of seedlings . For another side , mines and pits They favor the entry of pathogens . Severe damage as in the regions of Apatzingan , Michoacan; Sinaloa and the Lagunera Region cause defoliation and burning of fruits with reduction in performance and quality . If he dano shows up After fruit ripening , it considerably reduces the sugar concentration [2,17] .

Sampling and economic threshold . The economic threshold is not determined for this cultivation , but it is suggested follow the methodology recommended in tomato , which consists in place 30 x 38 centimeter plastic trays under the plants to capture larvae mature and what are you pupen in the trays , in time they do it in floor . The economic threshold with this methodology for the Southeastern California Coast in United States , is when there is an average of 10 pupae per tray by d^ a in 3 or 4 consecutive days [17] . When there are no pupae, although There are recent mines , it indicates that there is good natural control. If there is a percentage of parasitism greater than 50%, it is not necessary apply . A recommendation It is important not to stress the crop by lack of water during his development since this favors he miner increase [2.17] .

Control. Leafminer infestations at the beginning of the crop cycle are common , however , these are controlled by parasitoids . The use excessive insecticides against others pests , propitious he miner increase , because they are removed the parasitoids natives , so it is recommended sample to estimate the miner percentages and parasitism levels before any spraying . The first mines are detected in young leaves and this sampling can be perform , in addition to the trays , with traps yellow sticky , in order to determine the initial infestation and the leafminer species . Most of the insecticides recommended currently against this pest are highly toxic to natural enemies and pollinators such as dimethoate , diazinon, oxamyl and esfenvalerate . insecticides cyromazine , spinetoram and chlorantraniliprole possess a toxicity moderate or low for insects beneficial and are effective against leaf miner [2.15-17] .

Important pests secondary

fruit borer

This pest Fruit borer , hibernates as adult under leaves, grass or trash around the crop fields . Towards the end of May, the insect abandons its shelters and feeds on vegetation adjacent until the species horricola is established in field . The eggs last 4 to 5 days and are laid in groups of 3 to 10 on leaves or buttons floral . The larva requires 12 days at 35 degrees centigrade and 43 days at 23 degrees centigrade , passing by five urges . The pupal phase lasts 5 to 10 days . In addition to the melon, this plague it affects also to the "kabocha" pumpkin. The damage caused by the larvae is the flower stigmas , can undermine stems or petioles and feed on the leaves, interweaving them with silk threads . The larvae big prefer fruits in development , to which they drill presenting a fresh orange exudate . Upon penetrating the fruit , the larvae They seal the entrance with silk cloth .

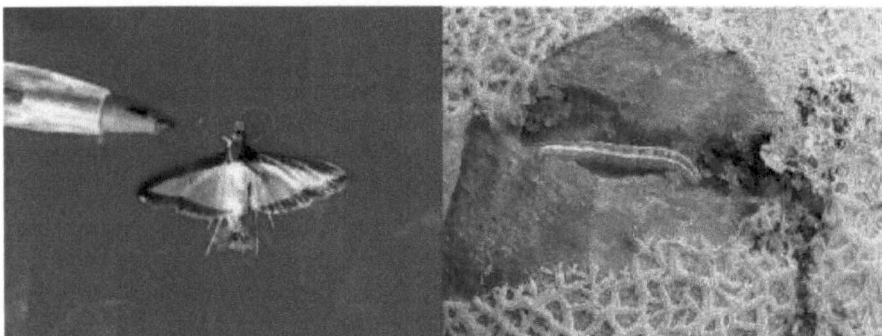

Adult and larva of Fruit Borer (Photographs Urbano Nava Camberos and Homero Sanchez Galvan)

It has been observed that the greater damage is recorded in melons cantaloupe type , being minimum in those of type smooth or *honeydew* . insecticides effective against pests and with toxicity moderate and low for pollinators and natural enemies are spinosad , spinetoram, methoxyfenozide , chlorantraniliprole , cyantraniliprole , flubendiamide and *Bacillus thuringiensis* [2,16,17] .

leafhopper green

The cicharrita green known also as Potato leafhopper or bean leafhopper is native to North America and attacks melon, alfalfa and others cucurbitaceae . They are metamorphosis insects incomplete going through the egg, nymph and adult stages . The eggs They last 8 to 9 days , the nymphs pass by five instars and require 8 to 14 days for their development before transforming in Adults . The population of this insect increases considerably in rainy conditions and presence of weeds , such as he quelite . Adults and nymphs They suck the sap from leaves, buds and petioles , injecting a toxic saliva that causes distortion of the leaves. in attacks severe produce chlorosis and necrosis of leaf edges , reducing the vigor of the plant. There are no reports of damage to the fruit . The insecticide acetamiprid is effective against this plague , possesses a toxicity moderate to bees and is partially selective to natural enemies . Others insecticides recommended for leafhopper control , such as diazinon , methomyl , esfenvalerate and imidacloprid are highly toxic to bees [2,17] .

Diabroticas

The diabrotics hibernate as Adults at the base of the plants , activating at temperatures of 18 to 22 degrees centigrade . They own metamorphosis complete going through the stages of egg, larva, pupa and adult . The egg has a duration of 5 to 8 days , the larva develops in he floor for a period of 15 to 30 days and the pupation requires 10 to 14 days . Adults feed on leaves and
flowers and in occasions they can ring the stems and foliar the seedlings

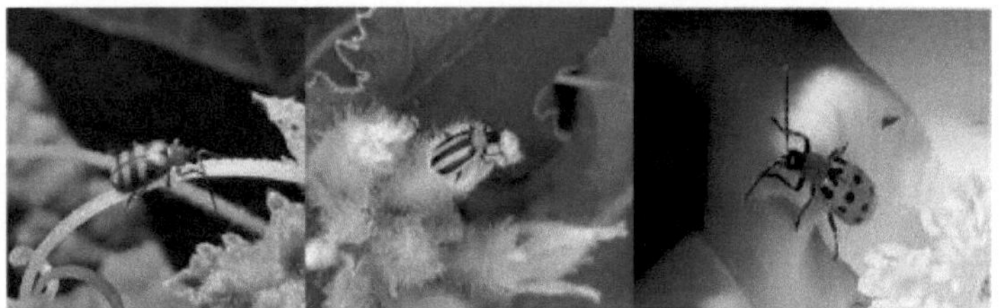

Diabrotic Adults (Photographs Urbano Nava Camberos and Homero Sanchez Galvan)

The larvae feed in the roots and base of the stems , reducing vigor or causing death . Feeding in the seedlings and when powerful winds , affect the plant population drastically , making necessary he replant . The most severe damage they are caused by adult by transmitting wilting bacterial in melon and cucumber. Also the diabetics disseminate pumpkin mosaic virus . The insecticide acetamiprid is effective against diabrotic diseases , it has a toxicity moderate to bees and is partially selective to natural enemies . Others insecticides recommended for its control such as diazinon, carbaryl and esfenvalerate are highly toxic to bees [2,6,16,17] .

armyworm

This species is a cucurbit pest spring and summer cycles on the Coast of Hermosillo and Comarca Lagunera . Has metamorphosis complete going through the stages of egg, larva, pupa and adult - moth -. The eggs are laid in masses , last from 3 to 5 days , later the larva passes for six instars , requiring 10 to 16 days and the pupation that occurs in he ground , lasts six days . Initially it feeds on the foliage of plants , however, its greatest damage is caused by in fruits , where makes holes irregular isolated or grouped . In most of the cases he Damage is superficial , affecting quality , occasionally developing inside the fruit . insecticides effective against armyworm , with toxicity moderate and low for bees and natural enemies , are spinosad , spinetoram, methoxyfenozide , chlorantraniliprole , cyantraniliprole , flubendiamide and *Bacillus thuringiensis* [2,6,16,17] .

Worm fake measurer

This pest has metamorphosis complete going through the stages of egg, larva, pupa and adult - moth -. Moths are habits nights that deposit eggs individually , the which hatch In 3 to 7 days , the larva phase passes for six instars , with a duration of 15 to 20 days , and the pupa lasts 6 to 12 days . Same as him armyworm , can appear both in Spring and summer melon plantings . The larvae feed on the foliage and high populations they can affect the seedlings , causing a delay in his development or its death , which causes a non- uniform harvest . In mature fruits , the larvae they can feed from the network, affecting his quality . insecticides effective against worm fake meter , with toxicity moderate and low for bees and natural enemies are spinosad , spinetoram, methoxyfenozide , chlorantraniliprole , cyantraniliprole , flubendiamide and *Bacillus thuringiensis* [2,6,16,17] .

Flea jumpy

This pest normally develops in floors wild during several months, there the larvae feed on the roots and when emerge the adults , these They invade the melon . The eggs are laid in he floor near the roots , hatching after 5 or 7 days . The larva lasts 14 to 28 days , and the pupa lasts 4 to 8 days . The damage caused because of the flea jumpy consists in holes acquaintances as " ammunition shot ". Is considered a plague primary in many floors as he tomato and importance secondary in the melon The insecticide acetamiprid is effective against fleas jumpy , it is moderately toxic to bees and is partially selective to natural enemies . Others insecticides recommended for flea control jumpy are highly toxic to bees such as malathion, carbaryl , lambda- cyhalothrin and clothianidin (2,17).

Grillo

Crickets are pests of melon in he autumn cycle , especially under irrigation systems pressurized , because it serves as shelter and as water source . They own metamorphosis incomplete going through the egg, nymph and adult stages . The eggs are laid in groups below the ground , those that hatch during he summer , nymphs pass by eight urge for a period of 50 to 80 days . The greatest damage is caused during the emergence of the crop , feeding on the stems , foliage and roots , which weaken until they die finally ; when appears in big populations , can to completely destroy the crop ; the greater damage They occur from the period from August to September . Others damage associated with these insects , are the presence of spots in the melon fruits caused by excrement , and occasionally damage to flowers by feeding which affects pollination . Old damage associated with crickets , they occur during the afternoon -night , because in he d^a they hide in cracks in he soil , weeds and water conduction pipes . The insecticide carbaryl applied as Cricket control bait does not pose a danger to bees . Others insecticides recommended for cricket control , such as malathion , carbaryl in spray and bifenthrin are highly toxic to bees [2,17] .

Red spider

The spider red is a plague potential of many cucurbitaceae , however, occasionally cause damage economical . Its life cycle has a duration of 6 to 8 days at a time temperature of 30 degrees centigrade , passing by egg , larva in three instars and adult . Females lay 4 to 6 eggs. by d^a for a period of up to one month . The infestation initial in melon it is for Adults taken by he wind of the crops neighbors . Adults and nymphs group together in colonies in he undersides of the leaves, where they feed sucking the sap , which gives it a tanned appearance to the leaves , reducing the amount of chlorophyll and capacity photosynthesis . in attacks severe , the leaves die , defoliating the plant. Damage is more common under high conditions . temperature in late spring. The mites they elaborate spiderweb on the leaves, which attracts dust , pretending foliage discolored The acaricides effective against this mite , with toxicity moderate and low for bees and natural enemies are dicofol, sulfur and bifenazate [2,17] .

Main melon diseases

The main Melon diseases are: 1) diseases fungus as drowning , vascular wilt , leaf blight and powdery mildew ; 2) diseases viral as he yellowing of the melon and various types of mosaics and 3) diseases caused by nematodes . In the Lagunera region , the incidence and severity of diseases varies according to the sowing date and stage . crop phenology causing losses in he yield and quality of the fruit [3,15] .

Drowning

The agents causal this disease are fungi that cause a minor plant density so it should reseed , causing a gap in he development of cultivation and harvest . The symptoms of drowning They begin with an injury at the base of the neck that progresses to strangulation , which causes wilting and death of the seedling . Drowning occurs generally in the crops early and incidence decreases in the crops intermediate and late . The fungicides Recommended for its control are Captan , Mefenoxam, Fluopicolide, Thiophanate. methico , azoxystrobin and metalxyl [3, 15, 17] .

Foliar rust

This sickness start with small circular appearance lesions watery that later turn dark brown surrounded by a green or yellowish halo . Are stains grow quickly , of 20 miKmeters or more in diameter , until covering the entire sheet. In them they are observed rings concentric dark , characteristics of the disease where exists a large production of spores that are dispersed by he wind and rain . Leaf blight can to provoke a defoliation severe , starting in the basal leaves , so the fruits remain exposed to the sun, which reduces the quality and quantity of commercial melon . The fungicides Recommended for its control are azoxystrobin , chlorothalonil , folpet and mancozeb [3,15,16] .

Vascular wilting

The organism that causes this disease is a fungus . The plants are infected in any development stage . The fungus is a soil inhabitant and penetrates the roots by natural openings or lesions , multiplying in he vascular system . When the infection start In the seedling stage , they frequently wither and die . In older plants , the symptom initial is a temporary wilting of one or several guides during the hottest hours during he day , and at night they can recuperate . The lower leaves become yellow and as the disease progresses , the Yellowing and wilting become more pronounced until the plant dies . In others cases there is a wilting sudden without leaf yellowing . Other characteristic of this disease is a crack or injury at the base of the stem light brown and later dark brown . In these lesions an exudate is detected rubbery When cutting transversely he stem , guides or petioles , are observed fabrics dead brown

When the fabrics they die it is observed on the surface of the themselves a growth white cottony that represents he fungus . The severity of this Disease is greater at soil temperatures between 18 and 25 degrees centigrade and decreases at 30 degrees centigrade . At higher temperatures , plants become infected . but they don't wither presenting yellowing and little development . The low Humidity of floor favors the pathogen and increases he wilting , as well as an excess of nitrogen , particularly in the form of ammonium . The application of the fungicide is recommended benomyl for the control of this illness , the use of varieties resistant melon and crop rotation [3,15,16] .

Cinderella

The causative agent is a fungus and causes serious damage in regions with climates warm and dry . This is because a Once the infection starts , the fungal mycelium continues to spread on the surface of the leaf regardless of the humidity conditions of the atmosphere . the cinder can infect severely to cultivation in a week . The optimal temperature is 20 to 27 degrees centigrade ; The infection occurs between 10 to 32 degrees centigrade . The first symptoms of the disease are detected in he underside of the lower leaves , where he fungus produces small appearance white spots powdery or milkweed composed of spores that emerge from the structures of the fungus . Are stains they can cover completely the leaf blade. Infected leaves become chlorotic , then brown or light gray and they die

Melon plants with symptoms of powdery mildew (Urbano Nava Camberos Photographs)

lack of foliage prevents he normal plant development and increases he " sunstrike " damage to the fruits . The mushroom also infects stems and stems youths . The fruits are smaller and misshapen and ripen prematurely ; Furthermore , he sugar content is reduced. In the Lagunera region , sowing dates intermediate and mainly the late ones , are the most affected by this disease . It is recommended he use of resistant melon varieties . The fungicides recommended for its control are sulfur , benomyl , chlorothalonil , cyflufenamid, kresoxim methyl , fluopyram + trifloxystrobin , myclobutanil, penthiopyrad, pyraclostrobin + boscalid , quinoxyfen, thiophanate metallic , triadimefon, trifloxystrobin and triflumizole [3,15-17].

Mosaics caused by virus

Cucurbits are susceptible to mosaic viruses in any development stage . When plants are infected between six to eight leaves , first symptoms are observed in the youngest leaves which show a mosaic pattern - yellow or light green areas alternating with green areas dark -, discolorations interveinal , rugae and leaf reduction and deformation , internodes become shorter and cases severe , older leaves they die When A plant is infected in the middle of the cycle , the guidelines existing ones are developed normally and produce healthy fruits [3].

The plants infected in earlier stages , produce few fruits of poor quality and they observe each other mottled or stained green and yellow , in addition , the fruits affected present shear resistance . In the Lagunera region there have been detected the following viruses: cucumber mosaic virus, watermelon mosaic virus , mosaic virus zucchini yellow and tobacco mosaic virus transmitted by aphids and pumpkin mosaic virus transmitted due to diabrotica [3], [15].

Yellowing and scab viruses of cucurbits

This viral disease was detected by first time in the Comarca lagoon in 1999 mainly in the crops late established July in forward and causing reductions of up to 50% in he melon yield . The greatest severity was observed in the areas of Paila , Parras, Valle de las Delicias and Laguna Seca in he state of Coahuila and in Ceballos, Durango and Jimenez, Chihuahua [18]. Currently , the melon yellowing is found distributed in the entire Lagunera region varying his incidence and severity in the melon producing areas . In 2006 it was reported in Caborca , Hermosillo Coast and Guaymas Valley, Sonora [19] . The symptoms They begin with a yellowing of the basal leaves that progresses gradually until it appears in the entire guide and throughout the entire plant and the fruit does not ripen . The vector is the fly white and not transmitted mechanically . In the Laguna Region in a May sowing , it was observed that the incidence was drops -12.7%- although hafra a high vector population and was detected after 47 days after sowing . In a August sowing , Symptoms of yellowing occurred 21 hours after sowing , although the mosquito population white adults and nymphs were less than in the May sowing , there was a 100% incidence of yellowing . This indicates that on the later sowing dates a large percentage of flies whites are carriers of the virus [3,15,20] .

Melon plants with symptoms of the Cucurbitaceae Yellowing and Stunting Virus (Photographies Urbano Nava Camberos)

Pesticide management selective for the protection of bees
Pesticides used in melon
Insecticides and acaricides recommended for pest control in the Comarca Lagunera and Costa de Hermosillo, Sonora, are: organochlorines such as endosulfan (Thiodan , Agrosulfan) and dicofol (AK-20, Kelthane); organophosphates as azinphos metallic (Gusation metHico , Guthion), malathion (Malathion, Lucathion), dimethoate (Rogor , Rotor, Roxion), diazinon (Diazinon), and parathion methyl (Folidol), monocrotophos (Azodrin), methamidophos (Tamaron), oxidemeton methyl (Metasystox), mevinphos (Mevinfos), naled (Dibrom , Selexone), ethion (Etion) and acephate (Orthene); carbamates as carbaryl (Sevin), methomyl (Lanate), and oxamyl (Vydate); and pyrethroids as fenvalerate (Belmark), deltamethrin (Decis), esfenvalerate (Halmark), fenpropatrin (Platino), and permethrin (Rostov, Ambush, Premier). These insecticides are characterized because it is wide spectrum , non- selective , neurotoxic , high toxicity to mammals , residuality short and contact action mainly , except he oxamyl with action systemic . They are also recommended insecticides unconventional synthetics , biorational and some organics such as nicotinoids : imidacloprid (Confidor , Citlalli), acetamiprid (Rescate) and thiamethoxam (Ripper, Actara); avermectins : abamectin (Agrimec); spinosins : spinosad (Tracer); diamides anthrax : fubendiamide (Belt); triazines : cyromazine (Trigard); diacylhydrazines : methoxyfenozide (Intrepid); acids tetronics or ketoenols : spirotetramat (Movento); botanicals : azadirachtin (Neemix); microbials : *Bacillus thuringiensis* (Dipel , Javelin) and *Beauveria bassiana* (Bea-Sin, Naturalis L) [15,16] .
The fungicides recommended for disease control fungi are: azoxystrobin (Amistar), elemental sulfur (Sagasul), benomyl (Benlate), boscalid + pyraclostrobin (Cabrio), tumban (Captan), carbendazim (Bavisttn), chlorothalonil (Bravo, Celeste), copper , folpet (Folpan), fosetyl-al (Fungial), kresoxim (Stroby), mancozeb (Flonex), metalaxyl (Tokat , Rolaxyl), myclobutanil (Rally), thiophanate metallic (Cercobin), triadimefon (Bayleton) and trifloxystrobin (Flint) [15, 16] .
Currently most of the insecticides conventional previously indicated are followed

using widely in he melon cultivation insecticides non- conventional , biorational and certified synthetics organic are used in minor degree currently for the control of the melon pest complex . Insecticide use varies greatly from region to region . other in our pa^s .

In the Lagunera region, 21 ingredients were used insecticide active in melon production during he cycle agricultural 2010, which corresponded to 44% of a total of 50 pesticides used . The most used insecticides were : endosulfan (60% of producers), carbofuran (58%), imidacloprid (47%) and methamidophos (42% of producers) and they made 1 to 4 applications of the insecticide endosulfan in sowing early and 2 to 9 applications in sowing late with doses that varied from 0.5 to 2.0 liters by hectare . in sowing early it was carried out a carbofuran application ; Meanwhile in sowing late, 1 to 4 applications were made , the doses used They were from 0.5 to 2.0 liters by hectare . The insecticide imidacloprid was applied 1 to 6 times. in sowing delays and doses ranged from 0.5 to 1.0 liter per hectare [21] .

In the Lagunera region, 25 ingredients were used active fungicides , some with action bactericidal and acaricidal , during he cycle agricultural 2010, representing 50 % of total pesticides used . The most used fungicides They were : chlorothalonil (58% of producers), metalaxyl -M (53%) and mancozeb (53%). From 2 to 10 applications of chlorothalonil were made at doses of 0.5 to 2.0 liters. by hectare , 4 applications of mancozeb at doses of 0.5 to 1.0 kilograms by hectare and 4 to 5 applications of metalaxyl -M at doses of 0.3 to 1.0 liters per hectare [21] .

Pesticides used in he melon cultivation in the Lagunera region , during he cycle agricultural Spring-Summer 2010 [21]

Pesticide type	chemical group	Common name (Ingredient asset)
Insecticide / acaricide	Organoclorado	Endosulfan
	Organofosforado	Clorpirifos etil, Dimetoato, Malation, Metamidofos
	Carbamato	Carbofuran
	Piretroide	Beta-ciflutrina, Cipermetrina, Lambda-cyhalotrina, Permetrina
	Formamidina	Amitraz
	Neonicotinoide	Acetamiprid, Imidacloprid, Thiametoxam
	Avermectina	Abamectina
	Espinosina	Spinosad, Spinetoram
	Triazine	Cyromazine
	Diamide antranHica	Chlorantraniliprole
	Botanical	Neem Extract, Garlic Extract

Pesticide toxicity

The classification of the toxicity of pesticides for mammals , including the humans , it is based in the LD 50 (Dose lethal causing 50 % mortality) expressed in milligrams by kilogram . In this case the DL 50 does reference to the obtained in rats when he pesticide is administered by v^ to oral or dermal in acute form . a concept Parallel is the acute LC 50 , which is the concentration of a substance in he air that causes the death of 50% of the test rat population ; is expressed in milligrams per cubic meter or in parts per million (ppm). The classification according to these criteria is presented in he following table [22] .

Classification of the pesticides based on his toxicity acute by v^ a oral for mammals expressed as DL 50 in milligrams by kilogram

Category	Band color	Solid more than	Until liquid	Until

toxicological				more than	
Yo Extremely toxic	Red	---	5	---	twenty
II Highly toxic	Yellow	5	50	twenty	200
III Moderately toxic	Blue	50	500	200	2000
IV Slightly toxic	Green	500	---	2000	---
Fungicide / bactericide	Benzimidazole	Benomyl , Carbendazim, Thiophanate metHico Propamocarb, Mancozeb			Thiabendazole ,
	carbamate	Cymoxanil			
	Cyanoacetamide	Chlorothalonil			
	Chloronitrile	Copper carboxHico			
	Copper organic	Iprodione			
	Dicarboximide	Azoxystrobin , Piraclostrobin			
	Strobilurin	Metalaxyl -M			
	Phenylamide	Captan			
	Phthalimide	Copper oxychloride , elemental sulfur , copper sulfate			
	Inorganic				
	Morpholine	Dimethomorph			
	Pyrimidine	Pyrimethanil			
	Quinoline	Quinoxifen			
	Triazole	Difenoconazole , Propiconozole , Tebuconazole			
	Antibiotic	Oxytetracycline , Streptomycin , Kasugamycin			

Based on DL 50 , the insecticides used in melon in the Comarca Lagunera you can classify as follows manner :

1. Extremely toxic : carbofuran (Furadan), methamidophos (Tamaron, Monitor)

2. Altamente toxicos : endosulfan (Thiodan , Agrosulfan), lambda- cyhalotrina (Karate, Kirio)

3. Moderadamente toxicos: abamectina (Agrimec), acetamiprid (Rescate), imidacloprid (Confidor, Rotaprid, Gaucho), clorpirifos etil (Lorsban, Clorver), permetrina (Ambush, Pounce), cipermetrina (Cymbush, Ammo, Cima)

4. Ligeramente toxicos: amitraz (Mitac), azadiractina (PHC Neem), spinosad (Tracer), spinetoram (Exalt, Palgus), cyromazina (Trigard), clorantraniliprol (Coragen), methoxifenozide (Intrepid).

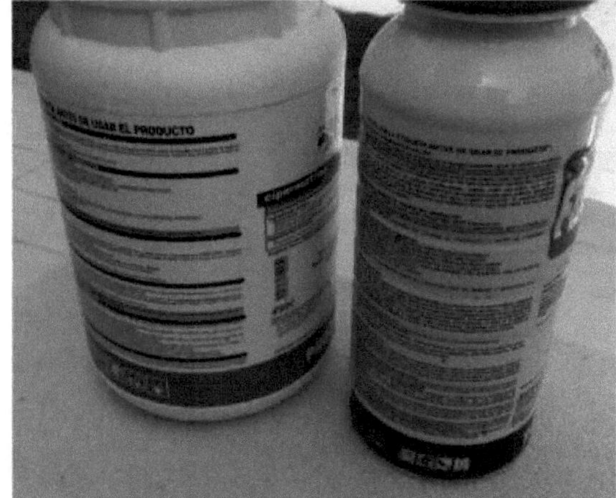

Pesticide containers with band green (slightly toxic) and blue (Highly toxic) (Photography Jose Luis Reyes Carrillo)

Classification of the pesticides based on his toxicity to bees

The classification of the toxicity of pesticides for bees , is based in the LD 50 , which is the dose lethal that causes 50% mortality of bees expressed in micrograms (pg) per bee you can notice in he following table [23,24] :

Classification of the pesticides based on his toxicity for bees , expressed <u>like DL 50 in micrograms per bee</u> [23,24]

Category	Toxicity	DL 50 Micrograms (pg) of ingredient active / bee)
Yo	Highly toxic	Less than 2
II	Moderately toxic	From 2 to 11
III	Relatively non -toxic	Greater than 11

The DL 50 in micrograms by bee you can convert to kilograms of ingredient asset by hectare of a given pesticide when sprayed about he cultivation , by multiplying by 1.12. For example , him carbofuran insecticide , which one is used widely by the melon producers from the Lagunera region , have an LD 50 of 0.149 micrograms by bee , the dose equivalent is 0.167 kilograms by hectare (0.149 x 1.12), which would be expected to kill 50% of the bees during his application . Considering that this insecticide was applied at doses of 0.5 to 2.0 liters by hectare (175 to 700 grams of ingredient asset by hectare) of the product formulated then the mortality it would cause would be very high , since with a dose relatively drops from 209 grams of ingredient asset by hectare (0.6 liters by hectare of product formulated) is estimated a bee mortality greater than 90 percent [23] .

insecticides used in melon in the Comarca Lagunera are classified as follows way based on his toxicity to bees [23,24] :

1. Altamente toxicos : clorpyrifos etil (Lorsban , Clorver), carbofuran (Furadan), permethrina (Ambush, Pounce), dimethoato (Rogor , Rotor, Versoato, Danadim), malathion (Lucathion , Cython), methamidophos (Tamaron, Monitor),

abamectin (Agrimec), cipermetrina (Cymbush , Ammo, Cima), lambda-cyhalotrina (Karate, Kirio)

2. Moderadamente toxicos: endosulfan (Thiodan, Agrosulfan), acetamiprid (Rescate), azadiractina (PHC Neem), spinosad (Tracer), spinetoram (Exalt, Palgus), cyromazina (Trigard)

3. Relativamente no toxicos: amitraz (mitac), clorantraniliprol (coragen).

Others products with use limited at the moment in melon and considered Relatively non - toxic to pollinators are: *Bacillus thuringiensis* (Dipel , Javelin), methoxyfenozide (Intrepid), tebufenozide (Confirm), diflubenzuron (Dimilin), flubendiamide (Belt), thiodicarb (Larvin), cyantraniliprol (Minecto Duo), spiromesifen (Oberon), buprofezin (Applaud), pyriproxifen (Knack), pymetrozine (Plenum), flonicamid (Beleaf), thiacloprid (Calypso), clothianidm (Clutch).

All the fungicides used in melon in the Comarca Lagunera are relatively non- toxic to bees [23,24] .

recommendations general for driving pesticide selective

For him design and implementation of a regional management strategy pesticide selective In melon, the following must be considered : recommendations :

1. Always alternate the groups insecticide chemicals to avoid he development of resistance and lack of effectiveness in the insects and mites

2. Preferably alternate conventional and biorational

3. Start with the insecticides biorational , for example Entrust, Dipel , Javelin, Crymax , Neemix , Trilogy, and PHC Neem

4. Continue with insecticides synthetics selective or low insect toxicity benefits and bees as Bt , Spinosines , Diacylhydrazines , Diamides , Acids tetronics , Sulfoxaflor, Emamectin Benzoate , Pimetrozine , Flonicamid , Dinotefuran , Acetamiprid, Chlorfenapyr .

5. Consider that the neonicotinoids (Acetamiprid, Confidor , Actara , Toretto) are very effective against insects suckers , selective for natural enemies such as predators and parasitoids , but toxic to bees . Preferably apply them to the irrigation system , not sprinkled

6. Use toward the end of the cycle the more toxic and harmful products for insects beneficial (organochlorines , organophosphates , carbamates and pyrethroids)

7. Consider that the following Insecticides only act against larvae (worms): Spinosad, Spinoteram , Emamectin Benzoate , Chlorfenapyr , Tebufenozide , Methoxifenozide and *Bacillus thuringiensis* .

Conclusions

The insects plague and diseases constitute one of the main limitations of the production and quality of melon cultivation , highlighting by his impact economical the fly white , powdery mildew and melon yellowing virus . Handling integrated pest and disease control of melon consists in the use of decision - making tools , such as sampling and monitoring , prediction and thresholds economical ; as^ as he compatible use of different control tactics or methods , such as cultural control, biological control , plant resistance and chemical control through the application of pesticides . At the moment Chemical control is the main control method used by the melon producers . In the Comarca Lagunera , up to 50 different pesticides during he melon production cycle . Main insecticides used by the producers are endosulfan (60%), carbofuran (58%), imidacloprid (47%) and methamidophos (42%); while the The most used fungicides are chlorothalonil (58%), metalaxyl -M (53%) and mancozeb (53%). Most of the insecticides at the moment used are toxic to humans , bees , insects pollinators native and natural enemies of insects plague . In general, the fungicides used possess low toxicity to pollinators ;

Therefore , it is necessary foment in the technicians and melon producers the implementation of a Management program Integrated Pests and Diseases with emphasis in the use of alternative control methods , such as cultural and biological control , plant resistance as well as the compatible integration of chemical control through pesticides selective and low toxicity to insects beneficial and the pollinators .

Pesticide spraying in he melon cultivation (Photography Olga Araceli Zapata Ramos)

References

1.   Nava CU, Rairurez DM. 2003. Management integrated pest control . *In:* 5°. Melonero Day . Techniques Updated to Produce Melon. Special Publication No. 49. INIFAP, North-Central Research Center, La Laguna Experimental Field. Matamoros, Coahuila. 2003;38-54.

2.   Rairurez DM, Nava CU, Fu CAA. Driving integrated pest in he melon cultivation . *In:* El Melon: Production and Marketing Technologies . Espinoza-Arellano, JdJ (ed.). CELALA-INIFAP. Mataroros , Coahuila. Technical Book . 2002;4:129 -59.

3.   Chew MY, Jimenez DF. Melon diseases . *In* : El Melon : Production and Marketing Technologies . Espinoza-Arellano JdJ (ed.). CELALA-INIFAP. Matamoros, Coahuila. Technical Book . 2002;4:161 -95.

4.   Sanchez GH, Cano RP, Avila GD, Rodriguez GL. Activities report , Campaign against mosquitoes silverleaf white , *Bemisia argentifolii* B. & P., in the Lagunera Region . Committee Coordinator of the Campaign against the White Fly , Secretary of Agriculture and Livestock . nineteen ninety six.

5.   Nava Cu, Cano RP, Martmez CJL. Driving integrated fly silverleaf white , *Bemisia argentifolii* Bellows & Perring. *In* : Garda GC, Medrano HR. (eds). Strategies for the control of vegetable pests , identification and control studies . COCYTED, SAGDR, CIIDIR-IPN Durango. Ed. Docu Imagen, Durango, Dgo . 2001;19-75.

6.   Ruiz CA, Bravo ME, Rairurez OG., Baez GAD, Alvarez CM, Ramos GJL, Nava CU, Byerly MF. Important Pests Economic in Mexico: Aspects of its Biology and Ecology . INIFAP-CIRPAC-Altos de Jalisco Center Experimental Field. Tepatitlan de Morelos, Jalisco, Mexico. Technical Book . 2013;2:447 p.

7.   Nava CU. Bionomics of *Bemisia argentifolii* Bellows & Perring on cotton, cantaloupe, and pepper. PhD Dissertation. Texas A&M University, Texas, USA. 1996;212p.

8.   Vera AM, Diaz PGR, Gonzalez CMM, Garzon TJA, Rivera BRF, Guevara GRG, Torres PI. Virus detection in tomato (*Solanum licopersicum*), chili (*Capsicum annuum*) and weed , in the different growing environments in Mexico (Advances) *In* : VIII Congress of Horticulture . Manzanillo, Colima, Mex.1991:132.

9.   Jimenez DF, Chew MYI, Cano RP, Nava CU. Etiology of yellowing of melon (*Cucumis melo* L.) in the Comarca Lagunera . *In* : Abstracts XXVII National Congress of Phytopathology , AC 9. Puerto Vallarta, Jalisco, Mexico. 2000;27.

10.   Butler GD, Henneberry TJ, Hutchison WD. Biology, sampling and population dynamics of *Bemisia tabaci* Agric Zool Rev. 1986;1:167 -95.

11.   Torres-Pacheco Y, Garzon-Tiznado JA, Brown JK, Becerra-Flores A, R. Rivera-Bustamante R. Detection and distribution of geminivirus in Mexico and the Southern United States. Phytopatology. 1996;11: 1186-92.

12.   Tonhasca A, Palumbo JC, Byrne DB. Distribution patterns of *Bemisia tabaci* (Homoptera: Aleyrodidae) in cantaloupe fields in Arizona. Environ Entomol. 1994;23:949-54.

13.   Palumbo, J. C., A. Tonhasca Jr. and D. N. Byrne. 1994. Sampling plans and action thresholds for whiteflies on spring melons. The University of Arizona, IPM Series No 1.

14.   Nava CU, Cano RP. 2000. Economic threshold for the midge white silver leaf in melon in the Comarca Lagunera , Mexico. Agroscience . 2000; 34:227-34.

15. Chew MYI, Reyes JI, Espinoza AJdJ , Rairurez DM, Pastor LFJ, Figueroa VU, Cano RP. Guide for Melon Production in the Lagunera Region.INIFAP , North-Central Regional Research Center , La Laguna Experimental Field. Matamoros, Coah . User information Technical . 2010;17:53 p.

16. Sabori P., R., J. Grageda G. and AA Fu C. Melon. *In* : Sonora Agricultural Technical Agenda. INIFAP. Mexico City. 2017;106-12.

17. University of California. UC IPM Pest management guidelines, Cucurbits.Publication 3445.2010 [En linea] (http://ipm.ucdavis.edu/PMG/selectnewpest.cucurbits.html) (Consulta 12/10/20).

18. Cano RP, Chew MYI, Chavez GF, Jimenez DF, Nava CU, Lopez RE, Avila GR, Castro IA. The yellowing of the melon (*Cucumis melo* L.) in North -Central Mexico. Possible causes and control strategies . Regional Plant Health Committee of the Lagunera Region of Coahuila and Durango. INIFAP-La Laguna Experimental Field. Torreon, Coahuila, Mexico. 1999;13p.

19. Moreno BA.Generalities and hosts of the cucurbit yellowing virus . *In* : Maldonado NLA, Fierros LGA. (eds). Management strategies integrated fly white and virus in cucurbitaceae . SAGARPA-INIFAP-CIRNO-Costa de Hermosillo Experimental Field. Hermosillo, Sonora, Mexico. Technical memory . 2007;26:37 -40

20. Nava CU, Chew M YI, Cano RP. 2007. Etiology , epidemiology and management of melon yellowing in the Comarca Lagunera . *In* : Maldonado NLA, Fierros LGA. (eds). Management strategies integrated fly white and virus in cucurbitaceae . SAGARPA-INIFAP- CIRNO-Costa de Hermosillo Experimental Field. Hermosillo, Sonora, Mexico. Technical memory . 2007;26:10 -28.

21. Vargas-Gonzalez G, Alvarez-Reyna, Guigon -Lopez C, Cano-Rtos P, Jimenez-Diaz F, Vasquez-Arroyo J, Garda-Carrillo M. Pattern of use of high- risk pesticides in he melon cultivation (*Cucumis melo* L.) in the Lagunera region . Ecosist Recur Agropec . 2016;3:367 -78.

22. SENASICA. Manual for the Good Use and Management of Pesticides on field. 1 [to] SAGADER Edition , SENASICA. 2019;76p. [Online] (https://www.gob.mx/senasica/documentos/manual-para-el-buen-uso- y-manejo-de-plaguicida-en-campo?state =published) (Consulted 10/19 /twenty).

23. Sanford MT. Protecting honey bees from pesticides. University of Florida. 2011;13p. [Online] (https://pesticidestewardship.org/wp-content/uploads/sites/4/2016/07/AA14500.pdf) (Consultation 10/19/20).

24. Hooven L, Sagili R, Johansen E. How to reduce bee poisoning from pesticides. A Pacific Northwest Extension Publication Oregon State University-University of Idaho-Washington State University.PNW 591. 2013;34p.[En linea] https://catalog.extension.oregonstate.edu/sites/catalog/files/project/pdf/pnw591.pdf(Consulta 19/10/20).

"If you kill a fly in March , you won't have to kill a thousand in May"
Saying Spanish

15. Weeds and competition with crop

Luis Enrique Moreno Alvarado, Eduardo Castro Martinez , Pedro Cano Rios and
Jose Luis Reyes Carrillo

Introduction

from latm *malitia* [1] , A plant is a " weed " if in any geographic area specifies their populations grow completely or predominantly in situations markedly disturbed by the man, not including by of course the plants deliberately cultivated . AsL the weeds include plants that are called wild , that enter on agricultural land , as well as those that are ruderal , this is from places wastelands and roadsides [2] .

weeds share some features , including [3] :

1   .- Long seeds life in he floor
2   . - Fast emergency
3   .- Ability to survive and prosper in the conditions disturbed from the field
4   .- Fast growth early
5   .- No requirements environmental special for the germination of their seeds 6.- They are competitive , often invasive and react in a similar way to cultivation practices .

Weed and Melon Floral Competition

The relationship between the presence of weeds and the Crop development is complex , because the abundance of weeds interferes with crop production .

The degradation and fragmentation of natural habitats is one of the main causes that affect negatively to the diversity and abundance of pollinators , however, the implementation of margins floral when offering sources pollen and nectar alternatives beyond the flowering of its own melon cultivation can improve the pollination services and , in consequently , improve harvests . In a study In the central part of Spain, both the floral coverage of the different species around us , such as their visits by pollinators to the melon, as well as he yield and quality of the fruit . As a result of the research, four species were identified. suitable to supply resources to pollinators : cilantro, yellow mustard , borage and calendula because they are species that received the greatest number of visits from pollinators since their blooms staggered they offered resources floral during several months in spring and summer . The composition of plants has to choose carefully , especially when the flowering of margins floral and crop they match . For example, it would be advisable avoid the concurrence of calendula flowering as It offers a large amount of pollen and nectar and would compete with the melon cultivation [4] .

Considering arid conditions in the lagoon region, it was studied he bee behavior during the pollination of the melon in spring since it was assumed that the flowering of the plants surroundings , already were cultivated or wild could compete advantageously by he pollen . Using he pollen hauled by the bees to the hive and captured with a trap for this purpose, it was found that his ID allowed determine the vegetation visited in that pollination period [5-7] .

The most visited plant species and it considered of greatest importance-in order of importance by he pollen volume captured - were he crop objective melon, the mesquite , alfalfa, governora, cucumber , mostacilla and sorghum respectively [8] .

Bee on mesquite flowers (Photography Juan Cabrera Reyes)

The plants visited in minor proportion They were maguey, trompillo , sand^ a , cuscuta , eucalyptus , quelite , ant grass , bindweed perennial , bindweed annual , yellow or tulip , ocotillo, corn , Tatalencho , swallow , Buffel grass , dandelion , sunflower chickpea , Chinese grass , stinky and wicker [8] .

foraging bee in the eucalyptus flowers (Photography Juan Cabrera Reyes)

With these results could be observe that the bees they visit primarily the melon flowers, but complement his feeding with plants blooming in the vicinity, cultivated , wild and weeds in he cultivation , indicating its paper also essential in pollination and seed production in his around .

Weed control

Mechanical control . The mechanical control includes the preparation of the field through plow or disc and cultivators . Mechanical control practices are among the oldest weed management techniques . Preparation of the seed bed through plow or disc exposes many weed seeds to variations in light, temperature and humidity . For some weeds , this process breaks weed seed dormancy , leading to early season control with herbicides or cultivation [9] . For the Lagunera Region , in he melon cultivation is recommended carry out weeding at 25 and 38 days after sowing , period in which still can enter the machinery [10] .

Manual control. Weed control through methods manuals is possible to do in the first melon development stages ; However, as the plant grows, the guides climb over the planting bed . thus limiting the use of machinery ; hence the common way of doing he weeding , after " closing " the bed is by hoe in irrigation ditch [11] . The number of weeds manuals during he Crop cycle varies between farmers . In the Laguna Region 22 % of melon producers carry out two to three , 46 % carry out four to six and only 14 % carry out seven or more weeds [12] .

Control with covers plastics . weeds compete and interfere with the crop in all type of agriculture but you want avoid weed control through products chemicals by several reasons [13] ; the most important are the demand for food free of pesticides , the evolution of resistance to herbicides in the weeds and issues environmental and health caused by the herbicides . Researchers from all over the world are developing non - chemical weed control techniques . Can prefer any method harmless control due to factors such as the nature of the crop , the characteristics ecological conditions of the area, the nature and intensity of weeds , the availability , effectiveness of other methods and factors social and economic [14] .

Laminated , irrigation by tape for weed control and use efficient water (Photography Olga Araceli Zapata Ramos)

The use of plastic to cover he floor this winning popularity well provides several benefits that are not limited to soil conservation , improved water efficiency , increased benefits economics , regulation of soil temperature and weed control . Only one is recommended black or occasionally colored plastic layer for your use in systems agricultural , since the plastic transparent is not as effective as the black. Below deck soil plasticity , it is known that the productivity of many crops , particularly vegetables as the melon increases significantly .

The black plastic cover inhibits strongly weeds when exercising a pressure physics about These block sunlight from reaching weeds or seeds , therefore inhibiting germination , and warming he floor causing a solarization impact . S^ the weed has already germinated or already this established suppress or at least decrease he growth . The use of a deck plastic also can cause stress by oxygen^gen due to exhaustion in he soil that would affect negatively the germination or weed growth [15] .

Cultivating he bed edge melonera padded with black plastic (Photography Olga Araceli Zapata Ramos)

The planting season recommended for Melon cultivation in the Lagunera region is from March 15 to April 15 ; However, in the region the planting period is extended from beginning of February until the end of May [12] . The sowing early february face he downside risk temperatures , which causes the seedling emergence period is delayed and they run he risk of being damaged by frost An alternative to produce planted melons in periods of sickness temperatures is the use of plastic tunnels that cover the planting row . In addition to frost protection , the use of tunnels can pass he start of harvest , increase the returns unitary and protect the crop from the presence of insects virus transmitters [10] .

Tunnel Agribon ® plastic for melon production (Photography Jose Luis Galarza Mendoza)

chemical control

The use of herbicides pre-emergent such as trifluralin and applied and incorporated into the soil through the passing of a Lilliston cultivator at the time of the second weeding are a alternative for the region and the following chart illustrates recommendation [10] .

Herbicide product	Commercial material / hectare	weed controlling	Form and time of application
Trifluralin	2 liters	Zacates annuals , purslane, quelites	Applies in weed preemergence when he melon have 3 to 5 true leaves , approximately 30 days after sowing and incorporate it with a cultivator

			before irrigation
Bensulide	10 liters	Zacates annuals , Quelite , purslane	Pre-sowing .
Sethoxydim	2 to 3 liters Add 2 liters of oil agricultural for greater effect	Grasses pinto, pegaropa , chino and Johnson (seed and rhizome)	Band. Applied after the emergence of the grasses and melon . The dose minor for grasses annuals and larger for perennials

Driving integrated

In general, in the crops weeds they can cause loss highest potential (34 %), being less Pests (18%) and pathogens (16%) are important . Weed control can be handled manually, mechanically or chemically , so the effectiveness is considerably greater than for the control of pests or diseases , which depend largely of products chemicals synthetics . Despite a clear increase in he use of pesticides , crop losses have not diminished significantly during the last 40 years . However the use of pesticides has allowed farmers Modify the production systems and increase the productivity of crops without suffering greater losses that are likely to occur by greater susceptibility to the effect harmful of pests . Crop losses are often minors are economically acceptable ; however, an increase in the productivity of crops without a protection appropriate of the crops make no sense [16] . Integrated weed control is the one that offers the greatest efficiency in melon and consists in he use of two or more methods . They are described below briefly some alternatives about he driving integrated weed in he melon cultivation [10] .

The regional melon producer practices he driving integrated weed when making use of the mechanical and manual control methods , based in the preparation of the land , sowing in humid to come earth , the use of cultivators to carry out weeding and the hoe for manual control . The first factor to consider in integrated weed control In melon it is the preparation of the planting bed , since the step of plowing , harrowing , edging , irrigation , covering the soil to the point of humidity with a Lilliston cultivator and sowing eliminate some generations of weed species in the first stages of development .

Other integrated weed control alternative In melon it is the preparation of the land and weeding associated with herbicide use selective Sethoxydim applied in post - emergence for grass control perennials as the Johnson and the Chinese. Weed perennial like grass amarosa , trompillo and coquillo are species difficult to control by media mechanical and chemical ; by Therefore, it is suggested that the manual control method be used by means of weeding and hoe .

Most important weeds

In the Lagunera region there is a wide range of weeds that invade he melon cultivation . There are a large number of weed species that are associated with the crop and whose Identification is essential for any control program [10] .

Cycle weeds annual . The species most common annuals associated with melon cultivation are the quelite , cadillo , bindweed , purslane, pinto grass, pegaropa grass and other minor ones importance , which are responsible for the competition by water , light and nutrients with cultivation 10 and [to] illustrate with images are weeds , the corresponding photographs and descriptions from the Pollen Atlas of the Comarca Lagunera [17] .

The quelite

Annual summer plant of 30 centimeters 1.8 meters high , with a thickness main stem with branches laterals , which are usually short ; leaves lanceolate or ovate , some times variegated , alternate and smooth from 5 to 20 centimeters long including the peduncles and 1.5 to 6 centimeters wide and present nails prominent veins whitish in he enves The flowers, which are not clearly visible , they are females and males, which are found in different plant in long spikes branched at the top of the [17th] floor . This plant invades land where it is planted the melon during spring, summer and autumn . Its characteristic of being a weed annual allows it to be fought by media mechanical and manual applied constantly . It is also found in ditch banks , roadsides and undisturbed areas . It is frequently used as fodder [10] .

The cadillo

Monoecious and annual plant with stem erect , robust up to 2 meters high , mainly simple and in occasions with branches at the base, covered with spots dark ; alternate leaves with long stalks , triangular blade up to 40 centimeters long by 3 to 20 centimeters wide; female flowers axillary in group of two, male flowers in heads globose located in clusters terminals or in the axils of the upper branches [17] . It infests the melon orchards of La Laguna during spring, summer and autumn. in percentages regular to severe . can be fought through weeding mechanics and weeding manuals . It is also found in roadsides , canal banks and little disturbed areas [10] .

The bindweed annual

Grass annual with stems voluble , simple, poorly branched , hairy , up to 5 meters long; leaves petiolate , alternate , heart -shaped , 3 to 10 centimeters long and 2 to 9 centimeters wide; flower axillary in groups of 2, with long peduncles ; corolla bell- shaped , purple, blue or red, 5 to 8 cm long and 5 centimeters in diameter . Flowering from July to November and with reproduction basically by seed [17] . It spreads in some areas where it is planted the melon His habit climber allows you entangle the plant, causing it give us the fruits and making it difficult his harvest . prevails during spring, summer and autumn and can be combated by media mechanical and manual [10] .

The purslane

Grass annual fleshy with stems prostrate or ascending , scattered radially ; leaves alternate , sessile , cuneate to spatulate , rounded or truncated in he green apex purple ; yellow flowers axillary in groups or solitary sessile ; fruit , a capsule 5 to 9 millimeters long with seeds circular black almost 1 miHmeter in diameter ; It flowers from May to November and reproduces by seed . This plant is used as human food [17] . Since it is a cycle weed annually you can fight satisfactorily by methods mechanical or manual . It is also common in temperate and tropical areas , in roadsides and little disturbed land [10] .

The pinto grass

It is a herbaceous plant that reaches 60 centimeters high with branches prostrate or ascending , knobby and with leaves 4 to 20 centimeters long and 3 to 8 millimeters wide. The inflorescences in 4 or more clusters of 1 to 2 centimeters in length of greenish or purple color [17] . In the Lagoon he is found widely distributed in infestations ranging from light to very severe . Presents itself in spring and autumn and causes reductions in he yield and quality of melon. It is also presented in crops that are planted in spring- summer , as well as in floors flooded , banks of ditches or canals and roadsides [10] .

The grass sticks

Annual plant from 10 to 90 centimeters . Leaves with Hgula ciliate and short , pods flattened , hairy on its margins . Inflorescence in panicle cyKndrica , sometimes interrupted at the base; rachis denticulate . mushrooms at the base of spikelets , with teeth retrorses , so that the inflorescence is rough when passed between the bottom fingers to above . Spikelets with 2 flowers, the upper one hermaphrodite ; the lower glume covers 1/3 of the spikelet [17] . He is observed in spring, summer and autumn . Is a grass annual that can be fight with weeds mechanical and manual . It is also found in curbs and edges of roads [10] .

Cycle weeds perennial

weeds perennials in he melon cultivation in the Lagunera region are Johnson grass , Chinese grass , grass bitter , the trompillo and coquille [10] .

Johnson Grass

Undergrowth perennial that has a fibrous root system , with rhizomes vigorous , resistant and penetrating , with purple spots and scales in knots . Erect stems with a thickness of 1.5 to 2 centimeters , cane - shaped , hollow , glabrous or finely pubescent in the knots . Height of 50 centimeters at 2 meters . Arranged leaves on two knees alternate along the stem , 10 to 50 centimeters long and 1.2 to 4 centimeters wide [17] . His combat by media mechanics and manuals is difficult and expensive . Invades areas where it is planted the melon but it is found in most of the spring summer crops as he corn , sorghum, cotton , grapevine, walnut and in irrigation canals , banks , fences , roadsides and little disturbed land [10] .

chinese grass

The leaves are green grayish , short , 4 to 5 centimeters in length , blades 0.5 to 6.5 centimeters long by 1 to 3.5 millimeters wide. The stems they can grow from 1 to 30 centimeters in height . The inflorescences They have 4 to 6 spikes , 1.5 to 6 centimeters long , and the spikelets 2 to 3 millimeters long . It has a very deep root system . The stems crawl by he soil and nodules they come out new roots , forming dense bushes [17] . In this region it is found during spring, summer and autumn . Its cycle is perennial and its forms of reproduction by seed and vegetatively make difficult and expensive his combat mechanical and manual. It is also found in cultivated areas , in roadsides ,

fences , irrigation canals and sites with poor drainage , as well as in little disturbed areas . Is very resistant to drought and alkaline soils [10]

Grass bitter

Perennial plant with stems erect, 30 to 60 centimeters high, with a green semi-woody base . some times bluish-green ; opposite leaves , short petiolate with lanceolate blade , 5 to 9 centimeters long with edge irregularly toothed ; flowers in flower heads 2 to 4 centimeters in diameter , peripheral flowers ligulate yellow in number from 14 to 20 and from 10 to 15 miKmeters . It is played by seed and underground stems [17] . Appears in canals , roadsides and poorly drained areas . This weed can be seen during all he year and because of his cycle biological perennial . Its mechanical and manual control is often fruitless and infests land planted with melon [10] .

Eltrompillo

Perennial erect one meter, simple branched stems at the top , fine pubescence and thorns yellow needles ; leaves alternate , petiolate , linear- oblong up to 15 centimeters and 5 to 30 millimeters wide with edge green wavy olive ; violet flowers , pedunculated , in tops scorpioid and long yellow stamens ; fruit , globose berry of 15 miKmeters , reproduces by seed and underground stems [17] . This plant is abundant in cultivated areas and in this region presents in any time of the year . Since it is a undergrowth perennial , its combat by media mechanics and manuals is difficult and expensive . It is also found in canals, roadsides and poorly drained areas [10] .

The nutsedge

Perennial plant with rhizomes thickened in the extremes called coquitos. erect stems triangular without branches neither pubescence and up to 90 centimeters high. Leaves only on the lower part of the stem , lanceolate , so long as he stem ; inflorescences in spikes of 8 to 20 panicles , flattened , golden yellow and 1 to 2.5 centimeters long. Each spike formed for 12 to 50 spikelets [17] . He is found regularly in infestations in melon Appears in crops vegetables , corn , cotton and in crops perennials such as alfalfa, walnut and grapevine. It is also common in meadows and gardens , roadsides , canal banks and in general in floors poorly drained wet [10] .

Impacto economic

In it Galia variety melon cultivation in a study under aridity and fertigation conditions , the total performance was reduced significantly only if he weeding began seven weeks after the emergency or later . The melon remained competitive and maintained performance potential optimum if kept free of weeds during the first month of crop growth ; he period critical for weed control in the melon was 4 to 6 weeks after the emergence of the seed , that is, the start of the fast weed growth and flowering peak male crop [18] . The studies indicate that according to the melon planting system in the Lagunera region , the presence of weeds annual start after 32 days after sowing with species as he quelite , purslane, pinto grass and pegaropa grass and they arrive to cause losses in production of between 30 and 40%. When they show up species perennials as Johnson grass , Chinese grass, grass bitter and trompillo he damage by competition is even greater [10] .

Conclusions

A plant is a " weed " if in any geographic area specific , its populations grow completely or predominantly in situations markedly disturbed by the man, not including by of course the plants deliberately cultivated . The bees they visit primarily the melon flowers, but also the plants blooming in the vicinity, cultivated , wild and weeds in he cultivation , indicating its paper also essential in pollination and seed production in his around . Weeds can be check by media mechanics , manual, covers plastics , chemical control and management integrated . The species most common annuals associated with melon cultivation are the quelite , cadillo , bindweed , purslane, pinto grass and pegaropa grass . weeds perennials are Johnson grass , Chinese grass , grass bitter , the trompillo and the nutsedge According to the melon planting system in the Lagunera region, it is estimated that the presence of weeds annual start after 32 days after sowing with species as he quelite , purslane, pinto grass and pegaropa grass and they arrive to cause losses in production of between 30 and 40%. When they show up species perennials as Johnson grass , Chinese grass, grass bitter and trompillo he damage by competition is even greater.

Severe weed invasion in bed melonera (Photography Olga Araceli Zapata Ramos)

References

1. Dictionary of the Royal Spanish Academy (DRAE). 2019. [Online] https://dle.rae.es/maleza?m=form (Consulted 10/24/20).
2. Baker H.G. The continuing evolution of weeds. Econ Bot. 1991;45:445 -9.
3. Zimdahl RL. Fundamentals of weed science. Academic press. London, UK 2018;664p.
4. Azpiazu C, Medina P, Adan A, Sanchez-Ramos I, del Estal P, Fereres A, et al. The role of annual flowering plant strips on a melon crop in Central Spain. Influence on pollinators and crop. Insects. 2020;11(1):66.
5. Reyes-Carrillo JL, Cano-Rtos P, Eischen F, Nava-Camberos U. Plant competition for honey bee pollinators during muskmelon bloom in La Laguna, Mexico. Proceedings of the American Association of Professional Apiculturists-American Bee Research Conference. Reno, Nevada, USA. Am Bee J. 2005;145(5):432.
6. Reyes-Carrillo JL, Cano-Rtos R, Eischen F. Plant Competition in pollination of cantaloupe (*Cucumis melo* L) with honeybees (*Apis mellifera* L.) in northern Mexico. Abstracts of the Apimondia Congress. Dublin, Irlanda. 2005;138.
7. Reyes-Carrillo JL, Munoz-Soto R, Colm-Cuevas L, Gaona-Gonzalez E. Pollen isolation and identification of wild, cultivated and ornamental plants in La Laguna District, Mexico. Abstracts of the Apimondia Congress. Dublm, Irlanda. 2005;141.
8. Reyes-Carrillo JL, Cano-Rtos R, Eischen F, Rodriguez-Martinez R, Nava-Camberos U. Plant species visited by honey bee foragers during induced cantaloupe pollination Acta Zoologica Mexicana. 2009;25(3):507-14.
9. Stall WM. Weed control in cucurbit crops (muskmelon, cucumber, squash, and watermelon). University of Florida, EDIS. 2009;3: 1-6
10. Cano-Rtos P, Espinoza-Arellano JdJ . The Melon: Production and Marketing Technolog ^as . Generalities of its production CELALA-CIRNOC-INIFAP, Mexico. 2002;4(1):1-18.
11. Moreno-Alvarado LE. Weed control with herbicides in melon in the Lagunera region . First day of the melon grower . SARH-INIFAP-CAELALA. Matamoros, Coahuila, Mexico. Special publication . 1990:33:1-2.
12. Espinoza-Arellano JJ. Situation of melon cultivation in the Lagunera region . Aspects technical and socioeconomic . First day of the melon grower . SARH-INIFAP-CAELALA. Matamoros, Coahuila, Mexico. Special publication . 1990;33:23 -35.
13. Chauhan BS. Seed germination ecology of feather lovegrass [*Eragrostis tenella* (L.) Beauv. Ex Roemer & JA Schultes]. PLoS One. 2013;8:e79398.
14. Hatcher P, Melander B. 2003. Combining physical, cultural and biological methods: prospects for integrated nonchemical weed management strategies. Weed Res. 2003;43:303-22.

15. Jabran K, Chauhan BS. Weed control using ground cover systems. Non-Chemical Weed Control. Academic Press. 2018;61-71.

16. Oerke EC. Crop losses to pests. J Agric Sci. 2006;144:31-43.

17. Reyes-Carrillo JL, Munoz-Soto R, Cano-Rfos P, Eischen FA, Blanco-Contreras E. Pollen atlas of the Comarca Lagunera , Mexico. Guzman Editores , Mexico, DF 2009;347p.

18. Nerson H. Weed competition in muskmelon and its effects on yield and fruit quality. Crop protection.1989;8(6):439-42.

"I had learned from the field a thing : that the best land is not seen because it is covered by weeds "

Juan Bosch

16. Nutrition of melon cultivation

Alejandro Moreno-Resendez, Pablo Preciado-Rangel and Jose Luis Reyes-Carrillo

Introduction

Inherent to the organisms existing on Earth, today d ^a you have knowledge relevant that have gradually permitted interpret photosynthesis as a process exclusively done by the vegetables , and breathing as a process common to all the beings alive . The evolution of knowledge about both processes , allowed elaborate the new nutrition model beings alive , which consists in a series of complex reactions that have level place cell phone , whose function is to provide the organism with energy and matter necessary to generate and regenerate its own structures . Therefore, the difference between plants and animals does not lie in the nutrition phase associated with breathing equal to but in the capacity of the first to synthesize products or compounds organic from substances inorganic , using the energy luminous , through the process acquaintance as " photosynthesis ". These knowledge have given rise to an irrefutable fact of nature , the vegetables are the responsible for the entry of energy into ecosystems 1 ·

In terms In general, plant nutrition is defined as he supply and absorption of compounds and or substances chemicals essential for the growth and the metabolism and items nutritious such as substances or compounds chemicals required by he crop in question [2] . Plant nutrition is a concept that must be drive to get results competitive within a production system , that is , decreasing losses and costs , maximizing efficiency and profits and obtaining high quality in the products generated within the agricultural sector [3] . The nutrition of the crop It is also important since it is related directly with the quality of the fruit in terms of size , appearance , texture , flavor , aroma, nutritional value and functional properties [4] .

Scary tractor he plastic edge in he mulch for planting the melon in beds (Photography Samuel Atahualpa Ramirez Contreras)

Nutrition and system soil -plant

The conditions optimal of the soils , will allow the plants perform in plant nutrition process . The combination of factors as he potential genetics of the plant and the stage of development , as well as the factors climatic - temperature , light, rainfall - and soil conditions - humidity , salinity , acidity , aeration - contribute to the differences in he growth and accumulation of matter dry , as well as the absorption and accumulation of nutritional elements [5,6] .

Items essential and dynamic in he system soil -plant

The elements nutrients used by species vegetables have different origin , as described below [7] :

1. Natural soil reserves : Soil composition , elements available and changeable -

the clays and matter organic , are the reserve source of soils - and conditions meteorological

2.   Fertilizers mineral or synthetic , a wide range of simple and compound fertilizers and microelements chelated and complexed and in less extent the fertilizers organic

3.   Irrigation water . Large amount of water circulates for the plants contributing mostly items such as calcium, magnesium , potassium , nitrates , sulfates boron

Mulching and watering by grooves in he melon cultivation (Photography Jose Luis Reyes Carrillo)

4.   Organic sources . Decomposition and mineralization of waste plants and animals of the soil . These They can be natural - recycling - or incorporated

5.   Rain. Especially nitrogen . rain water can catch and carry he nitrogen atmospheric towards the earth and join the system soil -plant

6.   Microorganisms : Fixation biological (nitrogen), mycorrhizae (phosphorus) and reactions of the items .

Forming part of the earth's crust there are a hundred elements natural chemicals , make up this cluster numerous items Several of them essential for the beings alive can to complete his Lifecycle . However , while some are essential for survival , the excess or the presence of others they can result toxic or even lethal [8,9] . The elements considered as Essential for all plants are 16: carbon (C), hydrogen (H), oxygen (O), nitrogen (N), phosphorus (P), potassium (K), calcium (Ca), magnesium (Mg), sulfur (S), iron (Fe), manganese (Mn), boron (B), zinc (Zn), copper (Cu), molybdenum (Mo) and chlorine (Cl). and four are only for some of these . All of them , when are present in quantities insufficient , they can reduce notably he growth and development of plant species [3,10] .

Others elements , such as he cobalt (Co), sodium (Na), nickel (Ni) and silicon (Si) promote he growth and may be essential only for some plants and are called beneficial , since both their concentration as his function are favorable although They vary between elements and plant species [8,10] .

mineral absorption

The absorption of items nutritious is carried out through the hairs radicals , the which during he plant activity period are in continuous renewal , given that its life lasts few days In conditions normal they can reach a quantity of 200 to 300 hairs root by miKmeter square , which is a large surface area for collecting elements . The absorption by unit length is maximum in the youngest areas of the root and decreases towards the most basal areas [3] . The absorption of large amounts of nutrients in periods short of time characterizes he request nutritional value of vegetables , including the melon, which is one of the most demanding cucurbits in relationship with fertilization, being he potassium he element most extracted from the soil [11] . The absorption of elements nutritious is different in the stages of crop development [12,13] .

Influential factors in mineral absorption

The absorption mechanism of items nutritious for the plants this related to your development status and conditions environmental , that is , it is affected by factors physiological internal as age , shape and potential plant genetics and factors external or environmental as he type of soil , radiation or energy light , temperature and environmental humidity [5] , [11] and equally for the practices cultural and soil conditions such as humidity , salinity , acidity , aeration and presence of toxic substances [14] . Since a wide number of factors that influence in the absorption of items nutrients are affected by the handling cultural , it is important meet them for a better decision making [3.7] .

Melon cultivation covering the entire bed melonera in full bloom (Photography : Olga Araceli Zapata Ramos)

Absorption of elements nutritious

To get a production adequate and a excellent quality of the fruits , it is necessary know the need nutritional and seasonal suitable for the fertilization of each plant species . Keep levels adequate fertility during the Development periods of high nutrient demand are important to optimize plant function 15 ·

An absorption curve is the graphic representation of the extraction of a nutrient and represents the amounts of this element extracted through the plant during his life cycle [16] . The use of element accumulation curves nutritious for crops , such as parameter for the recommendation of fertilization , is presented as a adequate indication of the need for these items in each stage of plant development [17] .

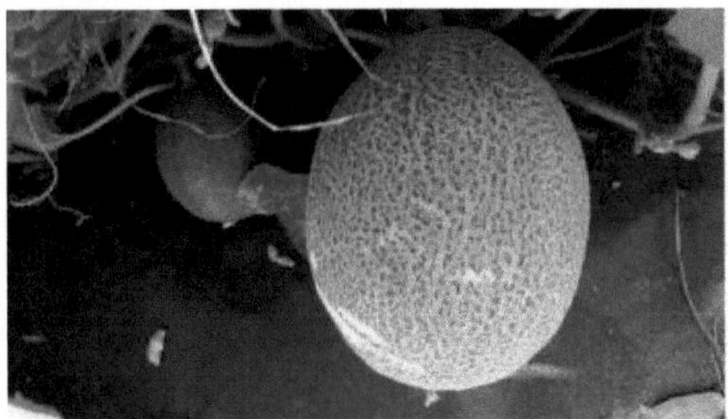

-type melon (Photography Olga Araceli Zapata Ramos)

To exemplify the significance of the absorption curves, we will use he study done by Mendoza-Cortez and collaborators [6] who they highlight that knowing he melon growth and development , nutrient extraction and distribution in the fabrics and in the different stages phenological characteristics of the plant, as well as the times of greatest demand for nutrients are information important that contribute to improving the planning and efficiency of fertilization of the crops .

These researchers used two melon cultivars , Iracema - type Yellow -, and Olimpic express - type Cantaloupe -, to evaluate he growth and accumulation of macroelements essential . They made sampling at 14 , 21, 28, 35, 42, 49 and 56 days after transplant (DDT).

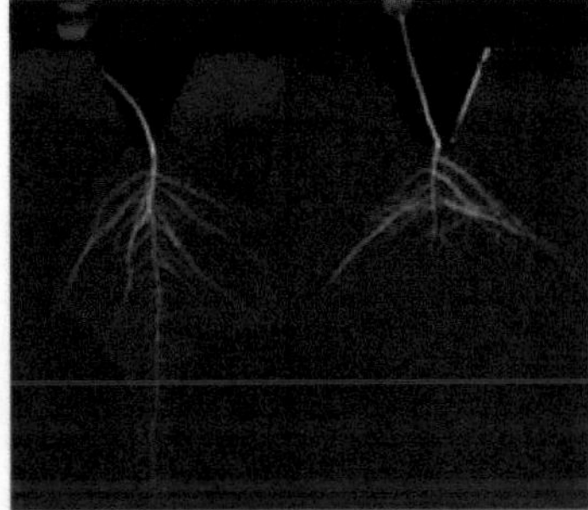

Root structure of *C. melo* . Left: triangular structure , predominant in the crops without phosphorus deficiency . Right : typical rectangular structure of deficiency crops nutritional phosphorus [18] .

Macronutrient accumulation

The production commercial and total of Olimpic express was 24 and 32 tons by hectare respectively , and in Iracema it was 30 and 38 tons by hectare . Olympic express accumulation a greater amount of macronutrients , especially N and K, characterizing a minor efficiency in he use of the nutrients for fruit production . The above is clearly seen in he following chart .

Accumulation of macroelements en them cultivars Olimpic express and Iracema, during he vegetative cycle [6] :

Element nutritious	Consumption	Olympic express	Iracema
Nitrogen		101.1	93.9
Phosphor		13.5	9.5
Potassium	kg/h a	173.4	136.0
Calcium		110.1	84.1
Magnesium		26.9	22.6
Sulfur		15.6	15.4

Derived from the results obtained, it was concluded that:

1. The plants of both melon cultivars have slow growth in the stage vegetative ,

189

intensifying the accumulation of Dry Matter during the stage reproductive , with a maximum accumulation of 246.4 milligrams per plant for Olimpic express and 266.9 milligrams per plant for Iracema, with participation of fruits of 60 and 64% respectively

2.   The majors macronutrient requirements They were between 28 to 56 days after transplanting and at the end of the cycle , the sequence of greatest accumulation of macronutrients of the cultivars evaluated continued he order decreasing in Olimpic express Potassium > Calcium > Nitrogen > Magnesium > Sulfur > Phosphorus and in Iracema Potassium > Nitrogen > Calcium > Magnesium > Sulfur > Phosphorus

3.   The amounts of Nitrogen , Phosphorus , Potassium , Calcium, Magnesium and Sulfur that were exported in the melon fruits , in relation to the totals accumulated , they corresponded to 61, 73, 66, 9, 35 and 39% in Olimpic express and 58, 70, 55, 6, 33 and 41% in Iracema.

4.   There is greater accumulation of Nitrogen , Phosphorus and Potassium in the fruits , while Calcium, Magnesium and Sulfur in the leaves of the melon cultivars , Olimpic express and Iracema

5.   With a minor fruit production and a greater accumulation of nutrients , the Olimpic express cultivar proved to be less efficient in he use of nutrients than the Iracema cultivar.

Ripe melons for harvest in Matamoros Coahuila (Photography Olga Araceli Zapata Ramos)

Discrepancy with conventional fertilization

On the last decades has increased society 's concern by he issue of food quality and, more in specifically , for the possible risks health that entails its consumption [19] . However the development of communities , communications , industry , and subsequent generation of energy and agriculture , particularly the highly technical and irrigation achieve growth in characteristics exponentials of the which the last decades are the most worrying [20] .

Today in d^ a is widely recognized that agriculture Conventional , even with the myriad of benefits it has generated over time , is a production system extremely artificial, based in the high consumption of inputs external -fossil fuels , agrochemicals and more- without considering the natural cycles , that is, it is the Production system agricultural in which they are used substances chemicals synthetically partial or total. Therefore, this system has also been responsible for various counterproductive effects [19] , [21] .

Clearing of huizache and mesquite trees for exploitation agricultural land in the Lagunera region (Photography Samuel Atahualpa Ramirez Macias)

Consequently , the concerns on soil degradation and sustainability agricultural they have woken up he interest in the evaluation of soil quality [22-26] . In it case of Mexico, the activities agricultural are the main cause

of the erosion of the soils , what it brings as consequence impacts negative on productivity [27,28] .

Laminated for melon cultivation (Photography Olga Araceli Zapata Ramos)

Alternatives to reduce he use of fertilizers synthetics

Dwarves recent , big and small producers , who traditionally they have used fertilizers synthetics to promote he development of their crops , they have modified this practice due to the restriction on he use of agrochemicals , the demand for high- quality foods quality and harmless , the concern growing due to resource degradation soil , the pressure of society by the aspects environmental , the savings and increased profits [29] . Additionally , the growing demand for food has established as alternative management sustainable of the production systems , promoting practices that preserve the natural resources and allow make a use efficient and adequate waste that is derived directly or indirectly from the agricultural sector [30] .

fresh soil cultivated in a melon farm in Matamoros; Coahuila (Photography Olga Araceli Zapata Ramos)

To decrease he fertilizer use synthetics , control diseases and parasites without the application of pesticides synthetics highly toxic and increase production are used biofertilizers because originate a fast decomposition of matter organic and assimilation of nutrients , consume little energy and do not pollute the environment [31] For this , it is necessary implement technologies that allow the application of these in the site and cultivation specific in order to fulfill its demand . In this sense , it has been pointed out that the use Nutrient efficiency is one aspect relevant , due to the increase in the costs and the impact environmental associated with his inappropriate use [26] .

The use of greenhouses and nutrition It represents a alternative to increase the production of crops , since it is a innovation that will allow speed up he change technological , which is one of the main causes of the low performances of different species vegetables . Furthermore , under production system protected can be control conditions environmental and reduce them infestations by diseases and pests [32] .

Melon cultivation in greenhouse tutored with raffia (Photography Alejandro Moreno Reséndez)

Application of biofertilizers and fertilizers organic in melon nutrition

It is important promote among producers of the Lagunera region he use of nutrition options for the melon cultivation . Like various strategies that have been implemented searching increase productivity agricultural , application of fertilizers organic , biofertilizers , bacteria plant growth promoters among he behavior friendly with him environment , the potential exploitation of the cycles biogeochemicals of the items nutritious and the least input dependence synthetics . Described below the results obtained by diverse investigations .

Application of vermicompost to melon . In a study in the Comarca Lagunera 33 it was determined he effect of four types of vermicompost (VC) on he melon cultivation . The vermicomposts were formulated for a period of three months with the action worm transformer *Eisenia fetida* from horse , goat , rabbit and bovine manure , and mixed with river sand (AR) with ratios 25:75 ; 30:70; 35:65 and 40:60 (VC:AR, percentage in volume); In addition , they were used pots filled only with sand such as control to which only solution was applied nutritious . The mixtures were placed in 20 kilogram black polyethylene bags in where they were planted Cantaloupe melon seeds .

The plants were led to a single stem , staking with raffia and the demand water was covered with irrigation by drip . The results indicated that in the mixtures VermiCompost: River Sand, with a 40:60 ratio (percentage in volume), regardless of the manures employees for your elaboration They were the best

.

The authors They concluded that not having used fertilizers synthetics during he crop development and fact that the melon cultivation will achieve to complete his cycle vegetative It allows assume that the different types of VC, due to their characteristics physics , chemistry and biology , they achieved satisfy the demand nutritious of this species and therefore the idea that vermicomposts have potential to support he development of the species vegetables , when used as part of the growth substrates .

At another job [28] where they were evaluated mixtures of vermicompost (VC) with sand (AR) [15:85, 30:70, 45:55, 60:40 (VC:AR, percentage in volume)] to determine his effect about he yield and quality of melon fruits cv. Cruiser developed under greenhouse conditions , it was concluded that the proportions 45:55 and 60:40 (VC:AR) presented an adequate nutritional status , as well as an increase in he performance (6.21 kg/m^2) in relation to the use of minors proportions (4.92 kg/m^2) and an increase of 27 percent hundred in the quality nutraceutical fruits . Consequently , the use of VC represents a viable alternative to use as source of nutrients and means of development for the melon cultivation in greenhouse , contributing to the preservation of the environment by reducing dependence on fertilizers synthetics .

Vermicompost-sand mixture for melon development in greenhouse (Photographs Alejandro Moreno Reséndez)

Application of biofertilizers to melon. In a study [31] of biofertilizers applied to the cultivation of reticulated melon variety ' Ovacion ' developed about padded with black polyethylene caliber 100 micrometers were used three biofertilizers commercials Z-Plex, Soil-Plex, Maya-Magic and a witness . It was evaluated he effect of the treatments about the fungus filamentous and mycorrhizal associated with the cultivation , the characteristics soil chemistry , yield and quality of the fruit . The authors concluded that the quantity and diversity of fungi , the factors soil chemicals , yield and fruit quality did not show effects for the application of the biofertilizers . Besides determined that the application of biofertilizers did not affect significantly the characteristics soil chemistry , fungus filamentous rhizosphere , nor he yield and quality of melon. The number of spores mycorrhizal increase 200 per cent with three biofertilizers , as well as he percentage of roots colonized from 12 by hundred in he witness up to 26, 30 and 48 per hundred for Maya -Magic, ZPlex and Soil-Plex. Finally They highlighted that although the biofertilizers presented potential to stimulate the presence and association of fungi - in quilted melon it was not achieved a symbiosis functional .

Melon seedling emerging from the black mulch (Photograna Olga Araceli Zapata Ramos)

Rhizobacteria . For study he inoculation effect bacteria plant growth promoters he melon behavior under greenhouse conditions [34] were used five strains of rhizobacteria and a control without application in variety melon seedlings Ovation developed in barren Peatmoss .

Thirty days After inoculation, it was transplanted into 20- liter pots , with tezontle as substrate and irrigation by solution drip nutritious . At harvest, the plant height , stem diameter , number and average weight of fruits were evaluated . per plant, yield , dry weight of roots and foliage . Bacterial inoculation did not increase he content relative chlorophyll , but if the plant height up to 50.4 per hundred when it was applied *Bacillus* cereus. With *Pseudomonas* fluorescens , the fruits with greater unit weight (891 grams). The dry weight of the root increase from 38 to 60 percent cent , the nitrogen and sodium in he foliage increased with inoculation and it was concluded that the use of bacteria in the production Melon hydroponics is a alternative to favor he cultivation development with practices friendly with him atmosphere .

Bioles . In a preliminary study [35] to determine he effect of the application of biofertilizers " Bioles " liquids - biopreparations crafts made from waste organic , rich in micronutrients , phytohormones and microorganisms benefits -, in doses of 5, 10, 15 and 25 per hundred in volume . They applied in sprinkles foliar weekly to see he development and severity of symptoms in melon infected with pumpkin mosaic virus . This concluded that the application of Biol at 25% in floors infected increase in 22 by cent the length of the plants and the number of leaves, like this as a reduction of 26 percent hundred in the severity of the symptoms provoked by the virus. It was found a high correlation positive doses applied of the bioles with the length of the plants and the number of leaves, like this as a strong correlation between the doses of the bioles and the decrease in symptoms provoked by the virus.

Application of various melon compost sources

Mushroom waste substrate . To study the mixture of waste from mushroom cultivation (SMS) with manure compost to obtain substrate and develop honey drop melon seedlings organically under greenhouse conditions The following were evaluated : 100 % SMS cent , mixtures of SMS and chicken manure compost, mixtures of SMS and livestock manure compost in different proportions : 80, 60, 40 and 20 per hundred . The results they obtained indicated that the addition of manure compost to the culture media based in SMS I provoke a decrease in alkalinity , an increase in porosity , an increase in conductivity electrical which is an indicator of salinity , the increase in the total water retention capacity and a favorable difference in nutrient concentrations . 80 per percent of SMS inhibited seed germination , however , the waste substrates from mushroom cultivation at 40 percent hundred generated the most appropriate condition for the seedling development [36] .

Compost from distillation residues . Under field conditions , compost derived from waste produced in the distillation cellar in a irrigated melon crop traditionally cultivated in the area where they are generated these wastes [37] ; were studied three compost dose : 7, 13 and 20 tons by hectare and one plot without applying compost. As a result , there were a improvement significant in he performance of the fruits in plots with application of 13 tons by hectare of compost and that compKa all the requirements to obtain returns elevated . Additionally, the application of compost improved the quality of the fruit with a higher solids content . soluble (°Brix). An effect of phosphorus was observed that caused an increase in he number of fruits individual in the plots that received the incorporation of compost.

waste compost solid urban . Searching reduce he use of traditional mobs *Sphagnum,* five culture media formulated from peat black , peat blonde , perlite and waste compost solid urban were used to obtain melon seedlings of the Eros cultivar, in a period of 25 days , without fertilization [38] . The authors established that waste compost solid urban It can be used as component in the formulation of mixtures with peat for the production of melon seedlings , providing a alternative ecological peat *Sphagnum* . The values elevated in salinity and pH found in this study appear as the main limiting factors his employment as cultivation substrate , although this can be solved doing compost mixtures with peat , which also would improve the properties substrate physics resulting . The mixture of Black Peat + waste compost solid urban + Phosphorus (65, 30 and 5% volume:volume , respectively) ensured the obtaining of melon seedlings with similar quality indices to those obtained with the mixtures conventional Black Peat and Blonde Peat. The salinity elevated can be corrected by washing appropriate substrate before utilization as a culture medium for plant production in pots without loss of production . waste compost solid urban provides not only a substitute cheap and high quality compared to peat but also a solution to the management of waste urban in the cities .

Application of fertilizers organic and inorganic fertilization to melon

Organic and inorganic to the soil and foliage . To determine the nutritional condition, development , yield and quality of the melon fruit with mulching and fertigation was evaluated your nutrition [39] through the application of: fertilization formula 180-100-200 (NPK), this same formula more activators organic and foliar inorganic , the same formula plus biofertilizers organic foliar and soil and finally the same formula plus hormones and inorganics to the foliage . There was an effect significant of the treatments in foliar Nitrogen and Phosphorus nutrition ; there was not differences in Total potassium and in Nitrogen . It was found a relationship between the amount of leaf nitrogen and potassium with the yield of Bruce quality melon (Jumbo). The use of foliar in fertigated melon favored he increase in the quality of the fruit and nutrition as well as the production and growth of the crop .

Manure compost and fertilizers . The application of manure composts and their combination with fertilizer synthetic on the properties of the soil and developed melon growth in pot under shade mesh conditions were studied during the autumn-winter [40] . The study was based in quantities similar total nitrogen in treatments equivalents . The treatments were application quantities for a livestock manure compost pot ; poultry manure compost ; fertilizer nitrogen , phosphorus and potassium (12-18-12); and one combination of both composts with fertilizer (12-18-12). The results showed that poultry manure compost favored a crop growth efficiency in comparison with livestock manure compost . Furthermore , the treatments fertilizer mixes synthetic and fertilizers organic obtained greater biomass of melon in comparison with simple compost treatments . Fertilizer treatment synthetic and poultry manure compost generated the best condition for him melon crop performance by improving the properties soil chemicals .

Conclusions

Melon nutrition is of great importance since it is related directly with the quality of the fruit in terms of size , appearance , texture , flavor , aroma, nutritional value and properties functional . Allowed three forms of plant nutrition : nutrition carbonated through the incorporation and transformation of carbon dioxide in carbohydrates in he photosynthesis process ; mineral nutrition, through the root assimilation of elements simple nutrients , and water nutrition This is the absorption of water for photosynthesis and with it the absorption of minerals . The production horttcola in systems protected is a alternative to production traditional in the field, especially in crops highly profitable . These systems they have proven obtain greater performance and quality , making use efficient of the nutrients and water . Various alternatives have been implemented searching increase productivity agricultural how to apply fertilizers organic , biofertilizers and bacteria plant growth promoters . Your application is favored by his behavior friendly with him environment , the potential exploitation of the cycles biogeochemicals of the items nutritious and the least dependence on supplies synthetics .

Driving and netting of melon fruit in greenhouse (Photography Alejandro Moreno Resendez)

References

1.   Gonzalez-Rodriguez C, Martinez-Losada C, Garcia-Barros S. The plant nutrition model throughout history and its importance for teaching . Rev Eureka Ensen Disclosure Science . 2014;11(1):2-12.

2.   Mengel K, Kirkby EA. Principles of plant nutrition. Ed. International Potash Institute. Basel, Switzerland. 5 th ed. 2001;849p.

3.   Research Institute Agricultural (INIA). Handling manual agronomic for melon (Cucumis melo L.) cultivation . INIA Bulletin N° 01. Santiago, Chile. 2017;92p. [Online] http://www.inia.cl/wp-content/uploads/ ProductionManuals /01%20Melon%20Manual.pdf (Query: 15/09/20).

4.   Luna-Fletes JA, Can- Chulim A, Cruz-Crespo E, Bugarin-Montoya R, Valdivia-Reynoso MG. Thinning intensity and solutions nutritious in cherry tomato quality . Rev Fitotec Mex. 2018;41(1):59-66.

5.   Contreras JI, Lao MT, Segura ML. Production and absorption of macroelements from melon cultivation in greenhouse under different NK dose and water salinity . Horticultural Rooordo . 2014;66:65-71.

6.   Mendoza-Cortez JW , Cecilio-Filho AB, Costa- Grangeiro L, Tavares-de Oliveira FH. Growth , macronutrient accumulation and production of cantaloupe and yellow melon . Rev Caatinga. 2014;27(3):72-82.

7.   Sanchez VJ. Soil fertility and mineral nutrition of plants - Concepts Basics -. s/f;1-19. [Online] http://www.exa.unne.edu.ar/biologia/fisiologia.vegetal/FERTILIDAD%20DEL%20SUELO%20Y%20N UTRICION.pdf (Consulted: 04/11/20).

8.   Azpilicueta C, Pena L, Gallego S. Metals and plants : between nutrition and toxicity . Magazine Science Today. 2010; 20(116): 12-6.

9.   Villegas-Torres OG, Dominguez-Patino ML, Martinez-Jaimes P, Aguilar-Cortes M. Copper and Nickel, microelements essential in plant nutrition . Rev Csc Nat & Agrop . 2015;2(2):285-95.

10.   Chen L, Liao H. Engineering crop nutrient efficiency for sustainable agriculture. J Integr Plant Biol. 2017;59(10):710-35.

11.   Aguiar-Neto P, Costa- Grangeiro L, Salviano-Mendes AM, Duarte-Costa N, Alves-da Cunha AP. Growth and nutrient accumulation in melon crop in Barauna-RN and Petrolina-PE. Rev Bras Frutic . 2014; 36(3):556-67.

12.   Hernandez-Diaz MI, Chailloux-Laffita M. Mineral nutrition and biofertilization in he tomato cultivation (Lycopersicon esculentum Mill.). Science and Technology Theme . 2011;5(13):11-27.

13.   Alarcon AL. Technology for high- yield crops . Ed. News Agricolas SA Murcia. Spain. 2000;460p.

14.   Rodriguez Z, Pire R. Extraction of N, P, K, Ca and Mg by melon plants (Cucumis melo L.) Hybrid Packstar under conditions of Tarabana , Lara state . Rev Fac Agron (LIGHT). 2004;21: 141-5

15.   Ku^ukyumuk Z, Ku^ukyumuk C, Erdal I. Seasonal changes of some plant nutrients under deficit irrigation programs of 'Braeburn' apple variety. J Food Agric Environ. 2013;11:(3&4):1687-91.

16.   Sancho VH. Nutrient absorption curves : importance and use in the fertilization programs . Agronomic   Information   s/ f;36:11 -13.   [Online]: http://intranet.exa.unne.edu.ar/biologia/fisiologia.vegetal/CURVAS%20DE%20ABSORCION%20DE% 20NUTRIENTES.pdf (Consulted: 04/08/20).

17.   Kano C, de Camargo-Carmello QA, da Silva-Cardoso S, Frizzone JA. Nutrient uptake by

greenhouse net melon. Semina: Cienc . Please 2010;31(1):1155-64.

18. Fita A, Nuez F, Pico B. Adaptation of the root system of melon (*Cucumis melo* L.) against deficiency in match . Agricultural Vergel. 2011;1(1):151-4.

19. Martmez -Castillo R. Soberania agri-food industry : characteristics , obstacles and perspectives , Science and Society. 2010;35(4):623-56.

20. Palacios-Velez OL, Escobar-Villagran BS. The sustainability of irrigated agriculture in the face of aquifer overexploitation . Tecnol Cienc Water. 2016; 7(2):5-16.

21. Lugo-Morin DR. Risk assessment agro - environmental community soils indigenous people of the state Anzoategui , Venezuela. Ecosystems 2007;26(1):69-79.

22. Kapoor J, Sharma S, Rana NK. Vermicomposting for organic waste management. Int J Recent Sci Res. 2015;6(12):7956-60.

23. Daza MC, D^az J , Aguirre E, Urrutia N. Effect of release fertilizers slow in nitrate leaching and nutrition nitrogenous in Stevia . Rev Colomb Cien Horric . 2015;9(1):112-23.

24. D^az-Franco A, Alvarado-Carrillo M, Alexander-Allende F, Ortiz-Chairez FE. Growth, nutrition and yield of squash with biological and mineral fertilization. Rev Int Environmental Cont . 2016;32(4):1-17.

25. D^az-Franco A, Alvarado-Carrillo M, Alexander-Allende F, Ortiz-Chairez FE. Growth , nutrition and yield of squash with fertilization biological and mineral. Rev Int Environmental Cont . 2016; 32(4): 445-5

26. Reyes-Perez JJ, Luna-Murillo RA, Reyes-Bermeo MR, Abasolo-Pacheco F, Espinosa- Cunuhay KA, Lopez-Bustamante RJ, et al. Use of worm castings and water hyacinth about he growth and development of cucumber (*Cucumissativus* , L). Biotecnia 2017;19(2):30-5.

27. Perez-Fernandez AR, Ruiz-Morales M, Lobato-Calleros MO, Perez-Valera E, Rodriguez-Salinas P. Substratum biophysical for agriculture protected and urban from compost and aggregates coming from the waste solid urban . Rev Int Contam Ambient. 2018;34(3):383-94.

28. Sanchez-Hernandez DJ, Fortis-Hernandez M, Esparza-Rivera JM, Rodriguez-Ortiz JC, de la Cruz-Lazaro E, Sanchez-Chavez E, et al. Use of vermicompost in the production of melon fruits and their quality nutraceutical . Interscience . 2016;41(3):213-17.

29. Fortis-Hernandez M, Sanchez-Tapia C, Preciado-Rangel P, Salazar-Sosa E, Segura-Castruita, MA, Orozco-Vidal JA, et al. Substrates organic treated for cucumber (*Cucumis sativus L.)* production under system protected . Hundred Tecnol Agriculture . 2013;1(2):1-7.

30. Hernandez-Rodriguez O, Hernandez- Tecorral A, Rivera-Figueroa C, Arras-Vota AM, Ojeda-Barrios D. Nutrient quality of four fertilizers organic produced from waste vegetables and livestock . Terra Latinoam . 2013;31(1):35-46.

31. Padilla E, Esqueda M, Sanchez A, Troncoso-Rojas R, Sanchez A. Effect of biofertilizers en melon cultivation with padding plastic . Rev Phytotec Mex. 2006;29(4):321-29.

32. Bautista-Hernandez CF. Effect of different nutritional sources on the productive potential of two varieties of pepper (*Capsicum annuum* L.) undergreenhouseconditions . Biotechnology 2017;19(1):17-2

33. Moreno-Resendez A, Garda-Gutierrez L, Cano-Rfos P, Martmez -Cueto V, Marquez-Hernandez C, Rodriguez-Dimas N. Development of melon (*Cucumis melo*) cultivation with vermicompost under greenhouse conditions . Ecosist Recur Agropec . 2014;1(8):163-73.

34. Rodriguez-Mendoza MN, San Miguel-Chavez R, Garda-Cue JL, Benavides-Mendoza A. Inoculation of growth-promoting bacteria in melon (*Cucumis melo*). Interscience . 2013;38(12):857-62.

35. Alvarez R, Espinoza L, Ruiz O, Peralta EL. Effect of the biofertilizers locally produced liquids " Bioles " , on he development of symptoms caused by Pumpkin Mosaic Virus (SqMV) in he melon cultivation (*Cucumis melo* L.) var. edited in greenhouse conditions . [Online]: http://www.dspace.espol.edu.ec/bitstream/123456789/17063/1/Robert%20Alvarez%20%20-%20Art%C3%ADculo%20Tesis%2014%20Sep%20Versi%C3%B3n%20Final.pdf (Consulta: 14/09/20).

36. Van-Tam N, Wang C. Use of spent mushroom substrate and manure compost for honeydew melon seedlings J Regul Crec Pl. 2015;34:417-24.

37. Villena R., Castellanos MT, Cartagena MC, Ribas F, Arce A, Cabello, et al. Winery distillery waste compost effect on the performance of melon crop under field conditions. Sci Agric. 20l8;75(6):494-503.

38. Herrera F, Castillo JE, Lopez-Bellido RJ, Lopez-Bellido L. Use of MSW Compost as alternative substrate to peat in melon seedbeds . 2019; Minutes No. 50. XI Conference of the Horticulture Group : 152-156 [Online] http://sech.info/ACTAS/Acta%20n%C2%BA%2050.%20XI%20Jornadas%20del%20Grupo%20de%20 Horticulture/Sesi%C3%B3n%20III/Utilizaci%C3%B3n%20del% 20Compost%20RSU%20as%20its treatment%20alternative%20a%20la%20peat%20in%20semilleros%20de%20mel%C3%B3n.pdf. (Consultation: 09/16/20).

39. Tapia-Vargas LM, Rico-Ponce HR, Vidales-Fernandez I, Larios-Guzman A, Pedraza-Santos ME, Herrera-Basurto J. Complementos nutritional for Performance and nutrition of melon cultivation with fertigation and mulching . Rev Mexicana Cien Agric. 2010;1(1):5-15.

40. Vo MH, Wang CH. Effects of manure composts and their combination within organic fertilizer on acid soil properties and the growth of muskmelon (*Cucumis melo* L.). Compost Sci Useful. 2015;23(2):117-27.

" Who pass cast , will have good harvest "

Anonymous

17. Protective and handling equipment

The team basic protection for checking hives as he beekeeper 's overalls , veil and gloves , smoker to calm the bees , cradle to open the hive and photographs that illustrate his usage are shown below

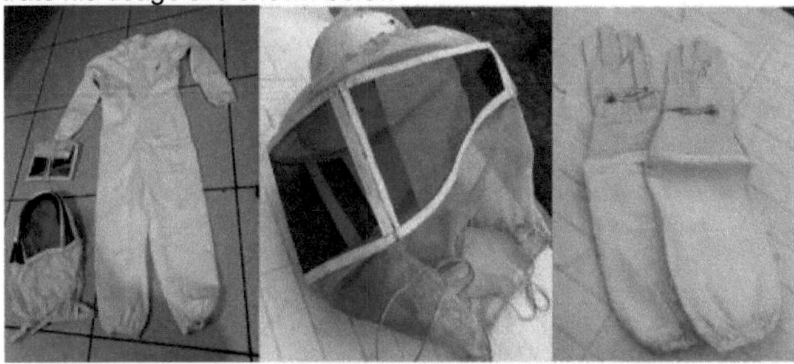

Overalls , square veil and beekeeper gloves

Smoker round , square and beekeeper 's cradles to open the hives

Beekeepers checking the hives with the smoker switched on showing round veils (photograph left), square and Sheriff type in the center (right)

jacket veil Sheriff type , far left) and square veil (photograph left) and square veil (photograph right) with smokers both round

Beekeepers with square veil , overalls , gloves , inspection equipment such as crib and smoker square with beehives jumbo type
(Photographs : Jose Luis Reyes Carrillo, Alicia Galeana Ramirez, Juan Cabrera Reyes and Hector Genaro Galindo Rodriguez)

"Only the bee will sting whoever " he handles it clumsily "

Saying Spanish

Credits photographs

Agustin Alberto Fu Castillo . Entomologist from INIFAP, Costa de Hermosillo Experimental Field. Email: fuca40@yahoo.com.mx
Alejandro Moreno Resendez. University Soil Department Autonomous Antonio Narro Agrarian , Laguna Unit. Email: alejamorsa@yahoo.com.mx
Alicia Galeana Ramirez: Agricultural businesswoman from Yecapixtla , Morelos. Email: maestria.ali.29@gmail.com
Fernando Morales Hinojosa. Beekeeper lagunero , bee producer and breeder queens . Email: honeymora@hotmail.es
Hector Genaro Galindo Rodriguez. Beekeeper from Hermosillo, Sonora. Email: hectorgalindo59@hotmail.com
Homero Sanchez Galvan. Entomologist at the Faculty of Sciences Biological - Juarez University of the State of Durango Email: sanchezgh@ujed.mx
Isabel Blanco Cervantes. Department of Biology , Autonomous University Agrarian Antonio Narro, UL. Email: miblancocervantes@gmail.com
Javier Cruz Nieto. Coordinated in Pronature Northwest AC Email: jcruzpictus@gmail.com
Jordan Hernandez Sanchez. Beekeeper from the state of Morelos. Email: apix-t@outlook.com
Jose Luis Galarza Mendoza. Technological Institute of Torreon, Anna municipality of Torreon, Coahuila. Email: galarzajl@yahoo.com.mx
Jose Luis Reyes Carrillo. Department of Biology , Autonomous University Agrarian Antonio Narro, UL. Email: jlreyes54@gmail.com
Jose Omar Enriquez Santacruz. Photographer magazine contributor Nomadica . Email: enriquezomar32@gmail.com
Juan Cabrera Reyes. Technological Institute of Torreon, Anna municipality of Torreon, Coahuila. Email: cabrera_reyes@yahoo.com.mx
Juan Ocon Cisneros . Beekeeper from San Pedro de las Colonias, Coahuila. Email: jocis_66@hotmail.com
Luda Marcial Salvador. Inter-American Institute for Cooperation on Agriculture (IICA), Saltillo, Coahuila. Email: lucy_leo11@hotmail.com
Mario Ruiz Caballero . Nature lover and amateur photographer . Email: maroruizcaballero23@gmail.com
Octavio Vazquez Calvete. Farmer , rancher and beekeeper Lagunero . Email: julyvazquez@hotmail.com
Olga Araceli Zapata Ramos. Graduate Student in Sciences Agrarian . autonomous University Agrarian Antonio Narro, UL. Email: aracelizapata-93@hotmail.com
Pedro Cano Rfos. Department of Horticulture , Autonomous University Agrarian Antonio Narro, UL. Email: canorp49@hotmail.com
Roberto Quintero Dominguez. Freelancer. Guadalajara Jalisco. Email: quintero.roberto@yahoo.com
Rodrigo Ramfrez Enriquez. Sales representative Idai Nature America SA de CV Email: ramzeu@hotmail.es
Romualdo Basilio Montiel. Baccalaureate Center Technological Agricultural # 104. Nazas , Durango. Email: romualdo.basilio@cbta104.edu.mx
Rubi Munoz Soto. Department of Biology , Autonomous University Agrarian Antonio Narro, UL. Email: rumuso23@hotmail.com
Samuel Atahualpa Ramirez Madas . Local Plant Health Board of the Lagunera Region . Email: samuelram2000@hotmail.com
Urban Nava Camberos . Faculty of Agriculture and Zootechnics-Faculty of

Sciences Biological , Juarez University of the State of Durango Email: nava_cu@hotmail.com

Vaughan Bryant. Texas A & M. University, College Station, Texas, USA. Email: vbryant@tamu.edu

Veronica Avila Rodriguez . Science Faculty Biological -Juarez University of the State of Durango Email: vavilar@gmail.com

Veronica Garda Mendoza. Business advisor agricultural Email: agronegociosrentablesmktg@gmail.com

Yasmm del Rodo Guevara Ramfrez . Veterinary Medical Student Zootechnist . autonomous University Agrarian Antonio Narro, UL. email : yazmin201997@gmail.com

"The life of the bee is like a well magical : the more you extract from it , the more it fills with water "

Karl Von Frisch

Glossary

TO
Bee, *Apis mellifera*
Bumblebees , *Bombus* spp
Tracheal mite , *Acarapis woodi*

mite , *Varroa destroyer*

Grooming , *grooming*
Aphelandra , *Aphelandra acanthus*
Avocado , *Persea Americana*
Drowning , caused by the fungus *Rhizoctonia solani* , *Pythium* spp and *Fusarium* spp
Garlic, *Allium sativum*

Ajonjoli , *Indian sesame*
Joy, *impatient crawling*

Alfalfa, *Medicago sativa*
Rise, boxes with their honeycombs that are placed over the hive for him honey warehouse .
Yellow tulip , Hibiscus coulteri
Stinky , *Sarcostemma cynanchoides*
Cranberry blue , *Vaccinium ashei*
Red spider , *Tetranychus* spp
Aster, *Leucosyris spinosa*
Wasps, Vespidae

b
Fruit borer , *Diaphania hyalinata*

Bendejo , *Hormathophylla spinosa*

Borage , *Borago officinalis*
Brezo , *Calluna vulgaris*
Broccoli , *Brassica oleracea*
C
Lizard flower , *Stapelia gigantea*
Folate , B Vitamin
Strawberry , *Fragaria* spp
g

Chickpea , *Peganum mexicanum*
Sunflower, *Helianthus annus*

Governor, *Larrea tridentata*

Governor, *Larrea divaricata* subsp. The Trident

Peanut , *Arachis hypogaea*
Cocoa, *Theobroma cocoa*
Columnar cacti ,
Pachycereus marginatus
Cadillo, Xanthium strumarium
Pumpkin "kabocha", *Cucurbita maxima*

Calendula, *Calendula officinalis*
Calliphoridae , flesh flies
Cantaloupe, *Muskmelon*
Cardenche , *Cylindropuntia imbricata*
Onion, Allium *cepa*
Cempoal , *Tagetes erected*
Cenicilla , *Podosphaera xanthi*
Cyclocephala , beetle family Scarabaeidae
Chicalote, *Mexican Argemone*
Chicharrita green , *Empoasca fabae*
Chile, *Capsicum annuum*
Cilantro, *Coriander sativum*
Hummingbirds , *Amazilia violaceps* , *Cynanthus sordidus* and *Cynanthus to the left*
Apodiformes, family Trochilidae
Copula, union sexual del macho y la hembra
Nutsedge , *Cyperus esculentus*
Corbfcula , leg pollen basket rear of the bee
Corniculos de los lice , a pair of appendages abdominal
Correhuela annual , *Ipomoea purpurea*
Correhuela perennial , *Convolvulus arvensis*
J
Jatropha , *Jatropha curcas*
Dark juno , *Dione juno*

I

Lantana, *Lantana camara*

Lavender, *Lavandula angustifolia*

Lettuce, *Lactuca sativa*
Lime, *Citrus limettioides*

vermicompost worms , *Eisenia's fetish*

Cosmopolitan , common to a great number of pa^ses
Cuscuta , *Cuscuta arvensis*
D
Circular dance, *round dance*

Wagle dance

Predators of the aphid , *Chrysoperla carnea* , *Hippodamia convergens* and parasitoids *Lysiphlebus testaceipes* and *Aphidius* spp
Diabrotica , *Diabrotica belteata* , *D. undecimpunctata y Acalyma trivitated*
Lion 's tooth , *Taraxacum officinale*

D^ptera , insects that have only two membranous wings

Drury , *Dysmorphia crisis*
E
Elzunia , *Elzunia humboldtius*
Finally pathogens effectives , *Beauveria bassiana* , *Paecilomyces fumosoroceus* , *P. farinosus* , *Verticillium lecanii* , *Metarhizium anisopliae y Aschersonia aleyrodis*
Sphinx hummingbird, *Macroglossum of stars*
Spermatheca , bolsa where he stores himself sperm
Eucalyptus , *Eucalyptus globules*

Exine , hard outer layer of the pollen grain
F

FABIS, *Fast Africanized Bee Identification System*

Families of bats specialized in flowers, *Phyllostomidae* and *Pteropodidae*
Felder, *Phoebe rurina*
Microgram (pg), millionimage part of 1 gram
Microsporidium, *Nosema* spp
Wicker , *Chilopsis linearis*
Leaf miner , *Liriomyza sativae* and *Liriomyza trifolii*

Whitefly , *Bemisia tabaci* biotype B = *B. Argentifiolii*
Flies commercial , Calliphoridae
Short- tongued flies , *Phoridae, Sciaridae* , *Mycetophilidae* and *Piophilidae*

Swallow , *Euphorbia micromera*
Cricket, *Gryllus (= Acheta)* spp
Grooming , grooming
Guanabana, *Annona muricata*
Worm false meter , *Trichoplusia is*

Soldier worm , *Spodoptera exigua*

H
Hediondilla, *Verbesine enceloids*
Hemolymph , liquid internal of the invertebrates ,
generally colorless , containing substances nutrients ,
although not oxygen

Hibernate, torpor during he winter .
Grass bittersa , *Helianthus ciliaris*

Ant grass , *Allionia incarnate*

Hymenoptera , arthropods membranous wings

Ants , Formicidae

Huizache , *Vachellia farnesian*

Yo
Instar , stage of development the insects to reach sexual maturity
Aphid parasitoids *Lysiphlebus testaceipes* and *Aphidius* spp
Parasitoides del minador , *Dyglyphusbegin , Solenotus intermedius* and *Chrysocharis* sp

Parasitoids natives , *Encarsia pergandiella , Eretmocerus teajanus , Encarsia luteola*
Pasto Buffel, *Cenchrus ciliaris*
Cucumber, *Cucumber sativa*
Pilladores , abejas thieves
Pinabete , *Tamarix* spp

Piquera , access to the hives so that the bees can Go in and go out
Pesticides microbials , *Bacillus thuringiensis* and *Beauveria bassiana*
Propolis , gum resinous than bees They collect from the buds of bushes or trees
Flea jumper , *Epitrix cucumeris*
Melon aphid , *Aphis gossypii*
Pupa, last stage of metamorphosis from larva to adult
Q

M
Macadamia, *Macadamia ternifolia*
Orange Magnolia , *Orange Magnolia*

Maguey, *Agave aspermia*
Ma^z , *The corn*
Majagua, *Hibiscus tiliaceus*

Mango, *Mangifera indica*
Apple, *Domestic malus*
Passion fruit , *Passiflora edulis*

Maranon , *Anacardium occidentale*

Wilting bacterial , *Erwina tracheiphila*

wilt , *Fusarium oxysporum* f. sp. melonis

long- winged butterfly , *Heliconius clysonymus*
Butterflies and moths , Lepidoptera

Monarch Butterfly, *Danaus plexippus*

Melon, *Cucumis melo*

Mesquite, *Prosopis juliflora*

American mesquite, *Parkinsonia aculeata*
Micron, thousandth part of a miKmeter

R
Rabanillo , *Raphanus raphanistrum*
Reticulated , with the appearance of red

Rhizobacteria , *Ochrobactrum anthropus , Microbacterium* spp ,
Bacillus cereus, Pseudomonas fluroescens and *Sphingomonas*
St
Watermelon, *Citrullus lanatus*
Syndrome , set of symptoms

Sorghum, *Sorghum vulgare* spp , various species

T
Tatalencho , *Gymnosperm glutinosum*
Tecomate , *Crescentia cujete*
Tizon foliar, *Alternaria cucumerina*
Tomato, *Solanum lycopersicum*
wheatgrass , *Phagopyrum esculentum*
Trumpet golden pine , *Angadenia berteroi*

Long- tongued flies , *Nemestrinidae* and *Tabanidae*
Mostacilla , bird chili , *Sisymbrium iron*
Mustard , *Sinapis alba*
mustard , *Diplotaxis virgata*
Bats , *Choeronycterus Mexican* and *Leptonycteris curasoae*
N
Turnip, *Brassica napus*
Nymph, similar to adult but in state immature
Walnut, *Carya illinoinensis*

Nosemiasis , protozoan microsporidium parasite
Nosema intestinal spp

EITHER
Ocotillo, *Fouqueria splendens*

Cat 's eye , *Pulicaria dysenterica*

Opercular, close with wax cell of a honeycomb

Q

Solitary palm, *Syagrus orinocensis*
Papaya, *Carica papaya*

Peat, light charcoal, spongy and with a earthy in places swampy due to the decomposition of remains vegetables
Peat , *Sphagnum*

V
Purslane, *Portulaca oleracea*
Fickle , very stems little endurance ,
Tobacco mosaic virus aphids , *Aphis gossypii, Macrospihum euphorbiae* and *Myzus persicae*

AND
Yucca, *Yucca* spp

Z
Zacate Buffalo, buffalo , *Cenchrus ciliaris*
Chinese zacate , bermuda , rooster leg , *Cynodon dactylon*
Zacate Johnson, *Sorghum* and pencils
Zacate sticky note , *Setaria vertically*
Zacate pinto, *Echinochloa column*
Carrot , *Daucus carota*

Chelite , *Amaranthus palmeri*   Trumpet , *Solanum eleagnifolium*   Buzzer cinnamon, *Selasphorus rufus*

> "A good one understanding few words"
>
> cardinal Mazarino

The bees melliferas and the pollination of the melon
Edition
Isidro Reyes Juarez
Style concealer
Robert Fifth Dominguez
Diseno
Isidro Reyes Juarez

Printed by Books on Demand GmbH, Norderstedt / Germany